PoliTO Springer Series

Springer, in cooperation with Politecnico di Torino, publishes the PoliTO Springer Series. This co-branded series of publications includes works by authors and volume editors mainly affiliated with Politecnico di Torino and covers academic and professional topics in the following areas: Mathematics and Statistics, Chemistry and Physical Sciences, Computer Science, All fields of Engineering. Interdisciplinary contributions combining the above areas are also welcome. The series will consist of lecture notes, research monographs, and briefs. Lectures notes are meant to provide quick information on research advances and may be based e.g. on summer schools or intensive courses on topics of current research, while SpringerBriefs are intended as concise summaries of cutting-edge research and its practical applications. The PoliTO Springer Series will promote international authorship, and addresses a global readership of scholars, students, researchers, professionals and policymakers.

THE SERIES IS INDEXED IN SCOPUS

Freddy R. Diaz Diaz

Architecture and Peacebuilding

Design Practices and Tools from Colombia

POLITECNICO DI TORINO

Springer

Freddy R. Diaz Diaz
Facultad de Arquitectura, Diseño y
Urbanismo
Universidad de La Salle
Bogotá, Colombia

ISSN 2509-6796 ISSN 2509-7024 (electronic)
PoliTO Springer Series
ISBN 978-3-032-13584-1 ISBN 978-3-032-13585-8 (eBook)
https://doi.org/10.1007/978-3-032-13585-8

This Springer imprint is published by the registered company Springer Nature Switzerland AG
The registered company address is: Gewerbestrasse 11, 6330 Cham, Switzerland

To the millions of Colombians who, regardless of which side, have suffered decades of armed conflict.

To all the people who still believe in peace and build it on an everyday basis.

Acknowledgments I would like to acknowledge my family for their unconditional support. To my tutors Prof. Paolo Mellano from Politecnico di Torino and Alex Leandro Perez from Universidad de La Salle de Colombia. To my coworkers at the Design for Transitions Laboratory of Universidad de La Salle de Colombia and to all those who have crossed my life on this path and have driven my passion for research and the contribution of architecture to a better society. To the peasants and ex-combatants of San José del Guaviare, Colombia, who received us with fraternal hospitality during the field work and who struggle daily to build peace.

Competing Interests The author has no competing interests to declare that are relevant to the content of this manuscript.

Contents

Chapter 1
Introduction

Abstract This chapter introduce the exploration of the monograph on the role of architectural design in Colombia's peacebuilding process following the 2016 Peace Agreement between the national government and the FARC-EP guerrilla. It examines the architectural responses developed within transitional frameworks such as the Development Programs with Territorial Focus (PDET) and the Territorial Spaces for Training and Reincorporation (ETCR), with particular attention to San José del Guaviare—a municipality heavily impacted by the armed conflict. The study adopts a multi-method qualitative approach grounded in participatory action research and structured in three phases: development of a theoretical model of architectural design for transition, analysis of built projects in the case study area, and formulation of design tools tailored to the Colombian post-conflict context. The research identifies the need to transcend conventional design paradigms by incorporating tools that respond to local complexities, promote community ownership, and foster autonomy. Through comparative analysis with similar international cases, the study proposes design methodologies that serve not only technical and spatial functions but also support social transformation and the non-repetition of violence. Ultimately, this work contributes to the creation of context-sensitive architectural practices that enhance the impact of built environments in peacebuilding efforts.

Colombia lived through an internal armed conflict of more than 60 years in which the State tried to stop the subversive advance of the FARC-EP guerrilla. Finally, in November 2016, a peace agreement was signed between the national government and this guerrilla, which opened the door to a peace-building process in which different local, national, and international sectors implement initiatives aimed at the reparation of victims, the reincorporation of former combatants and the non-repetition of the conflict. Those rural territories, marginalized and forgotten by the State, were the epicenter of the conflict and are today the scene of these transition processes in which spaces are needed to live, work, and meet.

On the one hand, the national government through Decree 893 of 2017 creates the Development Programs with Territorial Focus—PDET—as a planning and management instrument that has allowed the execution of more than 2,500 works of social

F. R. Diaz Diaz, *Architecture and Peacebuilding*, PoliTO Springer Series,
https://doi.org/10.1007/978-3-032-13585-8_1

and community facilities such as schools, sanitary facilities, and community centers. These programs have prioritized 170 municipalities mainly due to the impact of the armed conflict. On the other hand, the negotiation of the agreement also produced a transitory housing policy for the signatory combatants. These are the Territorial Spaces for Training and Reincorporation—ETCR -, 24 temporary rural settlements designed to accommodate the processes of preparation for the return to civilian life of the members of the FARC-EP. Likewise, international cooperation has been present through different organizations: the different UN agencies have been the most active, and have channeled resources through UNDP and FAO, supporting community projects for economic reincorporation, collective reparation, and food security.

The development and results of these projects managed by the different sectors are subject to permanent observation and verification. As of November 2022, six years after the signing of the Peace Agreement, the different architectural projects have produced some successful cases and others that have presented technical, structural, economic, and social difficulties, among others. Peace building has been as complex a task as the negotiation process itself. The role played by architecture in this process is relevant to the extent that the built space allows local communities to generate territorial appropriation, autonomy, and a vision of the future in order to seek the definitive overcoming of the conflict.

This monograph is an extended version of a doctoral thesis for the PhD in Architecture. History and Project of cycle 35 of the Politecnico di Torino which was recognized cum laude.[1] The book aims to review and understand the design processes developed in this context in order to identify successful practices and others to be improved or incorporated. This review is developed from the approach of a theoretical model fed by a discussion of theories and associated practices and the contrast with similar international cases. The main contribution of this monograph focuses on the proposal of architectural design tools that contribute to the architectural design practices necessary for the development of projects in the transition of a peace building process in a particular context such as Colombia. This research is based on the question: What architectural design tools can contribute to the transition process from war to peace in Colombia?

Architectural design in the framework of peace building in Colombia must overcome conventional paradigms and variables of architecture that have focused on solving the formal, spatial, technical, and functional problems of the project. The complex Colombian rural reality in which decades of armed conflict have developed requires a different understanding of local problems that are different in each place, generating design practices focused on the search for meaning and sense of architecture. In this way, it is possible to contribute to the appropriation of space, the vision of the future and community autonomy. The approach of architectural design tools focused on these practices is a useful and necessary dynamic device to improve the contribution that architects and designers can make in a context such as Colombia's in search of the definitive overcoming of the conflict and its non-repetition.

[1] The dissertation document is available at https://iris.polito.it/handle/11583/2983696.

The object of study defined for this research is the practice of architectural design and its role in the development of projects designed for the transition to peace in Colombia. This practice will be observed in the projects developed by different public entities, national and international cooperation institutions, and local communities between January 2017 and November 2021 in the municipality of San José del Guaviare.

San José del Guaviare is the 7th largest municipality in the country with an area of 16,178 km^2 [1]. It is defined as a case study because it was one of the most affected municipalities in the history of the conflict in Colombia due, among other reasons, to its geographic location at the entrance of the Amazon jungle, which provided a suitable scenario for the deployment of illicit activities by the different armed groups. During conflict 65,131 victimizing events such as attacks, forced displacements, evictions, among others took place (Unidad para la Atención y Reparación Integral a las Víctimas [2]). Likewise, San José del Guaviare is today home to important spatial dynamics around peacebuilding: two settlements of ex-combatants in reincorporation known as Territorial Spaces for Training and Reincorporation—ETCR. This makes it the only municipality hosting two of these 24 spaces located in the national territory, and 42 construction works of the Development Plan with a Territorial Approach PDET, the average being 14.7 constructions per municipality linked to the projects (Agencia de Renovación del Territorio [3]). Figure 1.1 shows the geopolitical location of San José del Guaviare in the national context.

With the aim of achieving the proposed objectives of this work, a multi-method research process is proposed, one that privileges the qualitative approach with an approximation to the methodological approach of action-participatory research, which suggests collaborative work between researchers and communities to attain social transformation. The work is structured in 3 chapters:

1. Development of a theoretical model that identifies the variables, parameters and indicators of architectural design projects associated with the transition in different contexts. Previously, a historical and spatial contextualization of the armed conflict in Colombia will be conducted. Subsequently, theories and practices associated with the concept of architectural design for transition will be reviewed in order to conceptually support the hypothesis. Additionally, a review of international reference cases that can show results in analogous scenarios of transition and social transformation in the global south will be carried out. Methodological tools such as documentary analysis, literature review, qualitative and quantitative analysis and comparative analysis will be used for this phase.

2. Analysis of projects in the area of study which allow to evaluate architectural design projects carried out within the framework of the implementation of the Peace Agreement, identifying relevant positive and negative aspects and practices that should be considered in the approach of the tools. For this purpose, the theoretical model produced in the first phase will be applied in the observation of projects located in the municipality of San José del Guaviare. The methodological instruments used for this phase are field work, participant observation, semi-structured interviews, life stories and qualitative and quantitative data analysis.

Fig. 1.1. Location of San José del Guaviare. By the author 2020

3. Proposal of architectural design tools for the transition that become an instrument for the generation of projects that effectively contribute to social transformation and peacebuilding in the national context. For this purpose, the tools will be previously classified according to their type of application and the task for which the tool is intended.

This work allows us to conclude, firstly, that the emerging responsibility of architecture in the face of the processes of ecological and social transition in the world implies rethinking the variables with which projects are traditionally conceived. Secondly, that architectural projects that intend to contribute to peace building in a context such as the Colombian one, must include localized design practices and the active and permanent participation of the benefited community, and finally, that in order to contribute to peace building from the field of architecture, dynamic and adaptable tools are required to facilitate design practices and contribute to the objectives set forth after the signing and implementation of the Peace Agreement.

References

1. Gobierno de Colombia (2022) Portal Único de Datos del Estado Colombiano. datos.gov.co: www.datos.gov.co
2. Unidad para la Atención y Reparación Integral a las Víctimas (2021) Registro Único de Victimas, https://www.unidadvictimas.gov.co/registro-unico-de-victimas-ruv/ .
3. Agencia de Renovación del Territorio (2022) Agencia de Renovación del Territorio, https://www.renovacionterritorio.gov.co/#/es/page/inicio

Chapter 2
Theoretical Model of Architectural Design for Transitions

Abstract This chapter presents a theoretical model made from the identification of variables, parameters and indicators related to the concept of architectural design for transitions. For this purpose, a spatial and historical context of the armed conflict in Colombia and the Peace Agreement is initially presented, useful for the localization of the discussion within the framework of the theoretical model application. Subsequently, a discussion of different theories associated with the concept of design for transitions in a global scenario of social and ecological crisis and the review of contributions from some architectural design practices emerging in the global south is proposed. Finally, there will be a review of two international reference cases that can show links between architecture and social transformation in peacebuilding scenarios. These inputs will allow the consolidation of the theoretical model that will be the basis for the Analysis procedure of the projects in the next methodological phase.

2.1 The Transition to Peace as a Scenario of Territorial Relations in Colombia

The armed conflict has been part of Colombia's history since the colonization of the Americas began at the end of the fifteenth century. Independence battles, inter-party struggles, border problems, drug trafficking and land struggles, among others, have made the conflict inherent to the history of our country in the last five centuries. However, this research intends to locate itself in "one of the longest internal conflicts in the history of the world" [9], the armed conflict with the guerrilla group called Fuerzas Armadas Revolucionarias de Colombia—Ejército del Pueblo, known by its acronym FARC-EP. This conflict and the period following the signing of the Agreement are framed in a complex socio-political and economic context that will be presented in the following paragraphs and that is intertwined with territorial, urban, and architectural spatial dynamics.

F. R. Diaz Diaz, *Architecture and Peacebuilding*, PoliTO Springer Series,
https://doi.org/10.1007/978-3-032-13585-8_2

2.1.1 Origins and Development of a Rural Armed Conflict

It is not possible to identify the exact moment of the beginning of the armed conflict with this guerrilla group because, although the foundation of this group was in 1964, it occurred after several years of violent events considered the basis and origin for its consolidation. The most relevant historical event in the period prior to the founding of the FARC-EP occurred on April 9th, 1948, when Jorge Eliécer Gaitán, a "charismatic liberal and leader of the agrarian reform movement was assassinated in Bogotá" [50]. This triggered a decade-long wave of inter-party violence known as "La Violencia", which claimed the lives of more than 300,000 Colombians.

The violence accelerated the colonization of unexplored territories in the eastern part of the country. Many land-working families from inland regions abandoned their territories to take refuge in peripheral areas such as the jungles of the plains and Amazonian foothills [63]. Thus, a first colonization process was consolidated in the remote areas of the Orinoquia and Amazon regions that "were promising for those who were looking for their own land where they could work and feel safe" [72].

During the 1950s and 1960s, settlers who came to occupy no-man's lands in the middle of the jungle and the plains founded hamlets and rural settlements with autonomous dynamics of spatial and territorial organization through which they delimited their lands and built their own houses and community services such as stores, canteens and river ports with the materials offered by the context. The actors of the conflict also gradually occupied places in these new settlements. [49] recounts in this regard that "By 1951, San José (del Guaviare), which at that time had no more than ten or twenty houses, a herd and very little agriculture, was invaded—so to speak—by guerrillas commanded by Palma and Morales, lieutenants of Alvaro Parra" [49].

At this time, the liberal and communist movements were positioned and strengthened in different areas of the country. In 1964, the army conducted a powerful armed attack against the so-called "Independent Republics" of two guerrilla commanders, and took over their camps [50]. This event, based on the accumulated progressive growth of the guerrilla groups, led to the integration that same year of the forces of 43 already active groups and the official birth of the FARC-EP.

Guerrilla movements in Colombia emerged in rural scenarios and as a response to a process of official violence: "political violence, violence as dispossession of land to make the boundaries of large estates grow" [25]. "They become articulated expressions of development efforts, as they seek to combat the incongruities and inconsistencies observed in today's society" [20]. The armed conflict consequently arose, remained, and was suffered mainly in the countryside, being this a relevant characteristic that differentiates it from other conflicts in the world that have had an impact on the cities. The struggle for land rights and tenure in rural areas in Latin America conditions the possibility of satisfying basic needs: food, housing, and work. "Judging by the analysis of historians, access to land is the main trigger of the conflict" [29].

A second wave of colonization was led by the State in the 1960s and 1970s again in some areas of the Orinoco and Amazon regions. This policy focused on granting plots of land so that settlers could produce basic foodstuffs, organized spatially in small villages with basic health and education services [55].

In total, 64% of the country's continental territory, which accounts for 737,454 km^2, corresponds to the Amazon and Orinoco regions. Since the time of Spanish colonization these territories were viewed with disinterest as they were not within the extractive network that privileged the territories located on the Magdalena River and therefore those connected to the ports of the Caribbean Sea [1]. This disinterest remained until the twentieth century, when the waves of colonization described above led to the growth of population centers and an increase in population. However, the centralist government proved ineffective in terms of the presence of state institutions that contributed to territorial development, not only in these two regions but also in the Pacific region in the west of the country and in those territories that went outside the central strip and the connection with the Caribbean Sea. As a result, this situation favored a continued marginalization that is still evident today, and the propitious scenario for the initiation of the armed conflict.

2.1.2 Consolidation of the People's Army and Crops for Illicit Use

In this context of marginality, the 1970s were a pivotal decade for the FARC-EP guerrilla, who went from being a movement of only about 500 people to an army of over 3000, thus consolidating a centralized hierarchical structure, a training school, and a political program [50]. During this period, the FARC-EP consolidated their influence, opened some new areas, and focused on training military leaders. Throughout the 1980s, they took control of many of the rural areas destined for coca cultivation as the main economic activity. To the settlers living in these areas "no legal crop offered the advantages that coca still has: the ease and economy of growing an Andean-Amazonian plant that does not need fertilizers or pesticides, a ready market" [50]. This bonanza generated a third wave of rural colonization of families who decided to devote themselves to these promising and profitable crops.

During those years, the economy associated with coca cultivation allowed for the daily movement of large sums of money that let the local inhabitants to cover their basic needs and, in some cases, even live ostentatiously. The 1980s and 1990s were lived in several regions of the country during socioeconomic dynamics controlled by the guerrilla who exercised several state functions in the absence of the State. Nicolas [18] defined from a legal approach "the other law" referring to La Macarena, a municipality located in the department of Meta: "Among the rural communities of La Macarena a series of laws and norms that regulate community life and define the social order are generally known" [18]. This "other law" allowed for coexistence between guerrillas and local inhabitants in several regions of the country.

The coca bonanza, as this ephemeral period of prosperity was known, generated a rapid growth of population centers. Commerce was the first sector to take a leap. Small grocery stores became large distribution warehouses [49]. Likewise, commercial activities that had not existed in those areas until then appeared, such as beauty parlors and movie theaters, along with others that were more common in a daily life revolving around an abundance of money, such as bars, taverns, and brothels. Churches and sports facilities also emerged during that period. “Real estate skyrocketed. All kinds of constructions were improvised, all kinds of small inns were set up” [49]. The FARC-EP expanded surprisingly in part because of the rise of coca as a source of financing through crops, laboratories, and alliances with landowners [77].

Another activity that favored the economic growth of the FARC-EP was extortion—collecting money through force or threats—which the guerrillas called “protection tax” or “association fee”. This practice led to an armed response from landowners and drug traffickers in the mid-1980s, which in turn, resulted in the creation of private self-defense armies known as paramilitary groups, which began armed activity as a counterinsurgency strategy on the fringes of the law. The growth of these groups and the declared fight against the guerrillas intensified the armed conflict. The paramilitaries, being outside the law, do not have the rules of combat that the army has as an official armed force, which is why the 1990s and 2000s were the most violent decades of the entire armed conflict, characterized by crossfire, massacres, forced displacements and other victimizing events in the search for territorial control in which the civilian population was caught in the middle. Tables 2.1, 2.2 and 2.3 present the most relevant historical aspects told from the different actors.

One of the attempts by a government to reach a peace agreement took place in 1999, when the government of the day created the Zone of Distension, an extensive area made up of five municipalities from which all its armed forces were withdrawn. The goal was to create the conditions for peace talks; however, they were not successful, on the contrary, the conditions allowed the FARC-EP to become

Table 2.1 History of the FARC-EP according to official sources. By the author, based on Colombia’s National Center of Historical Memory data, 2021

Period	Relevant historical fact
1964–1975	The emergence of the FARC-EP is related to the Army’s attacks against the so-called “independent republics”. There are confrontations between groups of different affiliations, such as the FARC against the ELN and the EPL
1975–1985	The FARC put a new rhythm to its gradual expansion: by the end of 1979 it already had nine fronts. The financing at the expense of the cultivation business generated internal friction since this conduct was contrary to revolutionary morality
1985–1995	Failure of the “Strategic Plan” which consisted of reaching 60,000 men and a conquest of urban areas. The FARC does not accept the conditions and becomes determined to intensify the war and defeat the State
1995–2005	The peace negotiations in the “clearance zone” collapsed under their own weight. The development of the Plan Colombia, the paramilitary expansion, and the policy of “democratic security” led to the military decline of the FARC

Table 2.2 History of the FARC-EP according to FARC-EP. By the author, 2021

Period	Relevant historical fact
1964–1975	Beginning of operation "Marquetalia" in response to the official aggression of a group of 48 small-scale farmers, in use of their right of defense. The "Southern Block" is constituted as the "Fuerzas Armadas Revolucionarias de Colombia FARC"
1975–1985	Consolidation and growth of guerrilla fronts. The Unión Patriótica party is born, and peace agreements are reached with the government of Belisario Betancour establishing a ceasefire
1985–1995	The movement joins the Coordinadora Guerrillera Simón Bolivar (CGSB), which unites several insurgent organizations. In conjunction with the CGSB, peace talks are established with Venezuela as an intermediary.
1995–2005	FARC-EP celebrates 34 years fighting for the dignity and future of the homeland. "The fascist Álvaro Uribe" presents the "Patriot Plan". This plan deploys a counterinsurgency force of 14,000 to 17,000 soldiers and mercenaries

stronger, especially in this part of the national territory. In 2002, the zone of distension was abolished and later that same year, the then incoming President Alvaro Uribe launched a government program called the Democratic Security Policy, thus increasing the defense budget and consequently strengthening the Armed Forces by regaining control of urban centers in scenarios of fierce and violent clashes. During the same period, there were numerous cases, now documented, of violations of the IHL (International Humanitarian Law) and extrajudicial executions by the Colombian Armed Forces and by State security agencies. The first decade of the twenty-first century is characterized by the intensification of the armed conflict in remote and marginal rural areas. "The country, with Uribe in charge, embarked on the most intense and harshest war option of recent times" [21].

In 2012, Juan Manuel Santos, two years after being elected President of Colombia, and Rodrigo Londoño Echeverri, commander of the General Staff of the FARC-EP, agreed to initiate the Peace Dialogues, setting up a negotiating table made up of delegates from the National Government, the FARC-EP and Cuba and Norway as guarantor countries. Four years lates, the dialogues resulted in the signing of the first peace agreement in the city of Cartagena in August 2016. After a renegotiation with some opponents of the Agreement, a new text for the Final Agreement was produced, which was signed on November 24th, 2016, in Bogota, and ratified by the Colombian Congress on November 30th of the same year. This event marks the beginning of an era called the post-agreement, the time and contextual framework of this research.

Table 2.3 History of the FARC-EP according to other sources. By the author, 2021

Source	Period	Relevant historical fact
Maria Alejandra Vélez Economist	1964–1975	The FARC were not protagonists at the national level, although they did consolidate themselves as social leaders who reestablished order in areas with little or no state presence
	1975–1985	The movement joins the Coordinadora Guerrillera Simón Bolivar (CGSB), which unites several insurgent organizations. In conjunction with the CGSB, peace talks are established with Venezuela as an intermediary.
	1985–1995	The FARC expands surprisingly in part because of the bonanza in coca financing. The 1990s continue to be the most favorable period for the growth of the organization
	1995–2005	The objective of urbanizing the conflict by locating in large municipalities close to roads to major cities seems to have been successful
Alfredo Molano Historian and writer	1964–1975	The army attacked the "Independent Republics" of Marulanda and Arenas and took over the camps. FARC consolidated its influence, opened some new areas, and concentrated on training military leaders
	1975–1985	The FARC went from being a movement of 500 people to a small army of 3,000, with a hierarchical structure, a general staff, a military code, training schools and political program. The government offers to negotiate peace
	1985–1995	The FARC-EP party, Unión Patriótica (UP) gains significant representation in the 1986 elections. President Ernesto Samper publicly acknowledges the political nature of the conflict
	1995–2005	With Plan Colombia and the Extradition Treaty, the U.S. subordinated public order and the development model to its interests. The FARC was forced to suspend the aspiration to develop a regular war strategy

2.2 Spatial Implications of the Conflict and the Agreement

The armed conflict was not only framed in a socio-political and economic context, but also conditioned the dynamics of spatial appropriation in the territory and in architectural development in rural areas. These decades of hostile rural scenarios allowed the emergence of empirical housing and equipment typologies, self-built by settlers that responded to daily activities such as licit and illicit agriculture, community meeting places and the demands of the local economy. Construction practices and materials also established an identity and a tradition as they emerged far from centralism and globalization and were influenced by the knowledge that the settlers brought into their new territories.

During the war, a large part of the national rural territory was the scene of armed confrontations that recurrently affected the civilian population and the spaces they inhabited, generating the destruction of villages, infrastructure, and communal equipment. According to the Unit for the Attention and Integral Reparation to Victims (2022), between January 1st, 1985, and May 31st, 2022, 11,905,917 victimizing events were registered, which have caused 1,060,493 homicides and 93,912 terrorist acts such as attacks and violent harrying [71]. The historical struggle for the right to "preserve productive property was repressed and supplanted by a hegemonic sector of agro-industrial, cattle and mining large landowners" [84].

The rural habitat became transitory and unstable due to victimizing events such as forced displacement and dispossession of land, which not only affected the patrimony of land-working families but also their roots, their appropriation of the territory and their physical and emotional well-being. From January 1985 to August 2021, a total of 9,218,796 people were victims of forced displacement [71]. The consequences are critical in both sending and receiving places where the increase in housing deficit and overpopulation of urban marginal areas is seen. "At all times the motive of the violent impulse is organically integrated to the expropriation of goods, especially land" [84].

Figure 2.1 shows a traditional rural dwelling in San José del Guaviare built of wood. During the war, its inhabitants were displaced, and the house was used for work related to coca leaf processing. Likewise, the combatants used strategies of ephemeral occupation of the space thinking about the dynamics of the war in which they had to be ready to react at any moment or move without leaving a trace. Figure 2.2 shows a platform called "caleta" that was used for sleeping. The excavation from which the earth was extracted served as protection in case of military attack. Before moving to another location, the hole was refilled with the same material.

The signing of the peace agreement also presents a scenario in which the socio-political and economic context is related to the territorial, urban, and architectural context. The laying down of arms and the end of the conflict agreed upon in the third point of the Agreement opens possibilities for rural communities to return to their land and to live and work on it, and from there contribute to the construction of peace. The first point of the Agreement agrees on a Comprehensive Rural Reform that recognizes the inequitable distribution of land as a problem to be solved through

Fig. 2.1 Traditional rural dwelling in San José del Guaviare. By the author, 2019

Fig. 2.2 "Caleta" or bedroom used by the FARC-EP combatants. By the author, 2019

the "democratization of access to land for the benefit of the rural communities most affected by poverty, state neglect and conflict" [23]. Likewise, this reform aims to "Contribute to the structural transformation of the countryside, closing the gaps between the countryside and the city, creating conditions of well-being and good living for the rural population" [30].

Currently the post-agreement is also the scenario of policies that derive in architectural projects. An instrument created by the Decree 893 of 2017, the Territorial Focused Development Plans (PDET), has allowed the execution of over 1500 construction works aimed at improving the community, educational, road and basic sanitation infrastructure [2] in 170 municipalities prioritized for having a high incidence of multidimensional poverty, a high incidence of armed conflict and a high concentration of coca crops [48].

The negotiation of the Agreement also produced a transitional housing policy for the guerrilla signatories. These are the Transitory Local Zones for Normalization where the first part of reincorporation was overcome and, as of August 15, 2017, the Territorial Spaces for Training and Reincorporation (ETCR) were established, temporary rural settlements intended to foster training and early reincorporation processes as to facilitate the adaptation of FARC-EP members to civilian life [3]. These two instruments, PDET and ETCR, generated new rural architectural typologies based mainly on prototypical and industrialized solutions. On the one hand, the community centers that have been built within the framework of the PDETs of which an example can be seen in Fig. 2.3, and on the other hand, models of rural housing for ETCR, such as the one shown in Fig. 2.4.

The scenario on which this research focuses highlights the relationships present in the Colombian rurality between the social, political, and economic context and the spatial context from the territorial, urban, and architectural scales. This implies not only the implementation of what was agreed in the Peace Agreement, with the support of international cooperation and the gradual positioning of state institutions that contribute to good living, but mainly the response from community organizations

Fig. 2.3 Community center in the rural area of San José del Guaviare. By the author, 2021

Fig. 2.4 Housing built at ETCR Charras, San José del Guaviare. By the author, 2018

that advance towards their autonomy. Carlos Velásquez, a peasant leader from San José, stated in an interview held in 2018 during fieldwork: "For the first time there is no armed group telling us what to do, and we do not know how to live without being told what to do. We must learn" [76]. To the extent that rural communities can remain on their land to work it, and with their basic needs met including housing, meeting places and communal productivity, progress can be made towards peacebuilding. "Whether or not peace is sustainable depends on its appropriation. The inhabitants are the key to the long-term construction of a peace infrastructure" [64].

2.3 Rurality as a Scenario of Conflict and Peacebuilding. A Contrast Between Europe and Colombia

As has been presented so far, the conflict in Colombia is a rural conflict and, therefore, the main scenario for action in peacebuilding is rurality. In this sense, in order to locate the discussions, tools and conclusions of this research, it is necessary to present some particularities of rurality in Colombia in contrast with the rurality in Europe and the global North.

During the 1960s, while in Colombia new rural territories were being colonized and those already consolidated were beginning to experience the armed conflict, the Common Agricultural Policy (CAP) was created in Europe in 1962, which represented a significant change in rural activity by proposing "a partnership between agriculture and society, between Europe and its farmers" [74]. The CAP from its inception and still today works to increase productivity and stabilize rural markets, ensure the availability of food at reasonable prices, and provide fair living standards for farmers. Additionally, it aims to contribute to the fight against climate change and the sustainable management of natural resources by keeping communities vibrant and preserving landscapes and rural areas throughout the European Union. "The CAP is a policy common to all EU countries. It is managed and financed at the European level through the resources from the EU budget" [74].

Parallel to the implementation of the CAP, changes in family farming emerged in Europe during the 1970s that gave rise to concepts such as "part-time farmers", "pluri-activity" and the "multifunctionality" of the family farm (multifunctional family farm) [38]. These processes, which emerge as a response to the development of capitalism and post-war industrialization in the United States and Europe mainly, "force family farms, especially those that do not have the economic means to buy more land, to diversify their sources of income" [38]. In part, these concepts are taken up by the approaches to the new rurality in Latin America in the first part of the twenty-first century. Edelmira [56] argues that the industrialization models of the European Community countries in the first decades of development also generated intense urbanization and massive depopulation of large rural areas: "In Europe, the rural sector continues to lose population due to factors such as soil depletion, changes in agricultural policies, urban concentration of economic activities, possibilities of collectivity and access to services, etc." [56].

It is important to highlight that violent forms of rural politics or activism were not exclusive to Colombia, nor are they, in general, a recent phenomenon, but rather, as Gerrit [35] argues, "a product of decades of formation and organization in the Global South." The catalysts for these outbreaks of violence were often agrarian unrest, the dispossession of indigenous lands, and changes in traditional patronage systems, among others, which were exploited by coercive entities with economic interests. These external agents evolved from urban movements scattered in rural geographies and/or caudillos who instigated insurgency. However, a specific event, an injustice or grievance was needed to incite the masses to take action, who, traditionally discreet and distrustful of the new or external, would follow someone who gained their trust to represent them or speak for them [35].

With generally violent and minimally transformative results, the rural Global South had dynamics of victim and victimizer in the struggle for agrarian power and had the advantage of an absent government. While these forms were commonplace, on the other hand, in the Global North, policies had existed for several decades that protected both rural and urban workers. In the United States, for example, national policies established since the Constitution, such as the National Labor Relations Act of 1935, served as a guarantee for union organization and collective protest actions aimed at combating negotiating power imbalances and ensured oversight processes in the face of injustices that persist to this day.

In this entire scenario of socio-political struggles, the importance of territory and the relationship with it is inherent for the peasant, indigenous person, or rural inhabitant, that is, ecology. The interaction with the natural environment is crucial for understanding the tense relationship between ways of life, the link between labor and capital, the overexploitation of resources, and the meaning of development. These conceptions, products of tradition or even spirituality, are now victims of political and scientific currents that "…have contributed to keeping social movements and ecological movements separate" [40] and are therefore inconsistent and unsustainable over time.

While organizations like the FAO in the Global South focused on re-educating people in basic ecological concepts through school manuals that covered topics such as human actions and their effects on the natural environment or the degradation of natural resources in rural areas, in the Global North, academia even theorizes about the very concept of nature and how "it has harmed us as a society to admire nature and agriculture as if they were on a pedestal" [52]. To this extent, it is clear how distant the inhabitants of the Global North can be from their natural environment as a means of subsistence, in contrast to the scenario faced by a large part of the Global South's population and how it uses rural areas as a stage for social development or simply exploitation.

Rural depopulation in Europe is being countered with incentives offered by some countries such as Italy or Spain to repopulate small municipalities in rural areas and with the "long-term vision for rural areas" adopted by the EU on June 30th, 2021. This is "a strategy to make Europe's rural areas stronger, connected, resilient and prosperous by 2040" [19]. These and other initiatives to strengthen rurality in Europe mean that a future of food shortages in that part of the world is not foreseeable. In contrast, other regions of the planet are experiencing a very rapid population growth, mainly in the cities, and growing depopulation in rural areas due to the lack of competitiveness of local agricultural products, social and economic instability, infrastructure deficits and, in countries such as Colombia, the inability of the State to establish itself effectively in these territories. This leads to increasingly important imbalances in the global food situation. "Therefore, developing countries will claim, with increasing force, their right to their share of the world's wealth, and it will become necessary for them to have the possibility of developing their own agricultural sector" [56].

Agricultural policies in Colombia have historically been defined by two models. The first, in the period from 1950 to 1990, known as the import substitution model, presented a structuralist approach that defined the causes of underdevelopment in Latin America in endogenous structural factors, responsible for an insufficient supply and, therefore, State intervention was fundamental to achieve development [78]. This approach allowed for the waves of colonization discussed above. The second model, known as the "economic openness model", emerged in the 1990s as a response to the debt crisis in Latin America. In Colombia, this model promoted the reduction of the size and intervention of the State, "under the imperative that demand and the free market should guide, as with an "invisible hand", the path of development" [78].

The crisis of the Colombian rural sector during the first years of the twenty-first century is also the result of the failure of the development model. In its National Human Development Report 2011, *Rural Colombia, Reasons for Hope* (2011) the United Nations Development Program (UNDP) highlights the negative results of the development model applied to rural Colombia in the last sixty years and the monumental debt that society owes to the rural world. The report also points to the lack of access to land and land tenure, the armed conflict and drug trafficking as causes of the crisis [59].

Transformations in the Colombian countryside have occurred in response to the failure of these models, to the violence, and to the presence of illicit crops. While in Europe the rural communities' welfare equates to the search to guarantee rural sustainability, in Colombia, "illicit crops are the unifying agent, as an economic opportunity to generate stable income, at the expense of the environment, the production systems and the rural social fabric" [53]. Other causes of the expansion of illicit crops are the decline in traditional agricultural production, the discouragement of the cultivation of domestic products, the dynamics of violent expulsion of the population and the cyclical crises of export crops. [61].

The present and future of the rural context in Colombia requires agrarian movements not only from state policies but also from grassroots initiatives arising in local communities that can boost productivity and generate viable alternatives to illicit crops. Overcoming the rural crisis requires the commitment of the whole society to settle the debt that the country has with the sector and to deploy its enormous productive potential [78]. This quest is assumed in the Comprehensive Rural Reform, the first point of the Peace Agreement. It proposes to "guarantee the development of the small-scale farmer and family-run economy as the basis of the country's food sovereignty" [23] as an alternative to solve rural poverty and the predominance of agribusiness and monocultures. Hence the importance of strengthening the countryside and the quality of life of small-scale farmers as a fundamental axis of peacebuilding. The effective implementation of the Accord is a historic opportunity to rectify the debts accumulated for decades with the country's rural population. Table 2.4 shows the conceptual synthesis of this section contextually locating the practice of design for transition.

2.4 Global Discussions on Transition and Action Scenarios for Design

Transition is an emerging concept that forces a critical review of the way we humans inhabit the planet. This implies reflecting on the identification of the role of design in the configuration of society and its different dimensions [15]. The following section presents a review of the theoretical discussions on global transition and the contributions from some localized practices. This discussion provides the framework for the generation of the theoretical model and allows an approach to the reflection on the role of architectural design in contexts of crisis and profound changes, and the responsibility of acting in processes such as peacebuilding in Colombia.

The world is going through an era of constant transformation framed in great ecological and social crises that lead us to the same destiny as a planet, but the outcome is still uncertain and controversial [60]. This historical moment has been

Table 2.4 Temporal context of the transition to peace. By the author, 2022

	Economic and socio-political context	Architectonic and urban territorial context
Past (War)	Rural colonization waves State centralism and marginality Consolidation of a populist army Attack to the established Positioning of illegal economies Military and political consolidation	Nomad and transitory rural habitat Infrastructure for war financing Loss of land ownership Precariousness of housing Deficit of common infrastructure
Present (Peace Agreement)	Surrender of weapons Return to former lands Presence of institutions in the rural areas Integral rural reform Reincorporation and reconciliation Productive projects	New human settlements – ETCR Infrastructure for productive projects Community social spaces – PDET New models for rural housing Improvement of tertiary roads Industrialized and prototypical solutions
Future (Peace)	Increase of political polarization Social and ecological crisis Rural self-governance Good living	Safe and sustainable rural habitat Ecotechnics applied to architecture Community and autonomous design processes Use of place-based materials Context based design

defined from different positions and critical discussions as the Ecozoic Era [7], the Great Transition [60], or simply the Transition [69, 36, 17] always having the planetary transformation at the core of the discussion and linking these crises with our way of inhabiting the planet and with the model of social life that has become dominant in recent centuries, categorized as industrialism, capitalism, modernity, (neo)liberalism, anthropocentrism, among others [17].

In this context, the concept of transition is not simply a transformation from one state to another, but instead it relates to a complex set of reactions in response to the need for change, a deep crisis or the obsolescence of a lifestyle; a profound change in response to the planetary challenges of the time [60]. There have been other transitions at critical moments in history, such as the colonization of America or the industrial revolution. In a major transition, the structure of society is exposed to a transformation of its cultural components and its relationship with nature [60]. The transition we are living in the world refers to the culmination of a rapid change resulting from different historical events of the twentieth century such as the world wars dependence on oil, the establishment of the United Nations and the emergence of the Internet, and our behavior in this historical moment will determine the features with which we will emerge as humanity in a new state [60].

The transformation required by this transition era must permeate all sectors of our daily lives and our social structure. James [69] defines five transformation scenarios to achieve the transition towards a sustainable society: demography, technology, economy, social equity, and governance [69]. Paul [60] includes in his Great Transition Initiative, seven driving forces or trends that are conditioning and modifying the system: demography, economy, social aspects, culture, technology, environment, and government. These dimensions reinforce and amplify each other in an accelerating process of transformation [60] that requires the unleashing of enormous amounts of creative work and inventive practices that will enable us to survive and flourish on a planet that is getting warmer [80]. Table 2.5 presents a summary of the dimensions discussed and the action scenarios evidenced where designers for transition can operate and contribute to the common goal.

In the possible scenarios, design then becomes a daily tool for social, cultural, and economic configuration and transformation at the local scale because of its action

Table 2.5 Action scenarios facing the dimensions of global transition. By the author, 2021

Dimension	Action scenario
Demography	Stability of the population
	Effective coverage of basic utilities
	Integral conditions for all demographic groups
	Response to population density and migratory flows
Industry and technology	Agricultural innovation
	Access to land and land tenure
	Energy transition
	Industrial process transformation
Economy	Circular and local economy processes
	Waste reduction and recovery
	Market regulation
Social equity	Sustainable lifestyles
	Fair distribution of benefits and economic obligations
	Assistance to the population in conditions of vulnerability
	Integral conditions for youth and women
Governance	Investment and financing of science, technology, and innovation
	Generation of public policies in response to the crisis
	Valuing traditional knowledge
Environment	Environmental impact reduction and control
	Environment and biodiversity care
	Water care

orientation. According to Thomas [7], in the Ecozoic era we live in a transition in which humans must be present on the Earth with the aim of mutually enhancing one other and becoming functional participants of the planet. Doing this involves "reinventing what is human" [7]. The reconfiguration must be immediate, as the rapidity of planetary transition increases the urgency to achieve a vision and to act, without which "we will cross thresholds that will irreversibly reduce options [...]. Delaying the adjustment of the way we live together on this planet may frustrate the opportunity for a Great Transition" [60]. In this regard, philosopher, and design theorist Tony [24] argues that "it can now be said that there is no survival without reorientation, and this imperative is the measure of the importance of design" [24].

Although these theoretical positions of transition are of global interest, there is a difference in how practical performance is understood and prioritized in various parts of the world. In the global north, actions with respect to transition have focused on responses to the ecological crisis. Countries such as Italy, France and Spain have created ministries for ecological transition covering sectors such as environment, energy transition and sustainable development. In contrast, in the global south, the social crisis is more evident and appears as a priority on government agendas before ecological issues.

"Recent experiences shape an increasing number of multilateral exchange relationships between projects of ecological transformation and participation operating in the North motivated by crises -of development-, and projects operating in the South motivated by crises -of induced underdevelopment-" [44].

Transition is understood from the epistemologies of the South as the processes of integral transformation supported by a structural change as a society that can reduce the social gap that exists in our peoples. Social equity and the search for a fair distribution of economic and environmental benefits, both within and between countries, is necessary for the transition to a sustainable society [69]. It is in this transitional framework where the peacebuilding process in Colombia takes place since land and economic resource tenure have been frequent causes of armed conflicts. In this sense, the transition that overcomes war and conflict is part of the transition towards sustainability [60]. Figure 2.5 shows a graphical synthesis of the readings of global territorial transitions.

Southern epistemologies present endemic positions on transition based on socioeconomic quests that make them different from some of the possibilities of those of the North. Some of these include ideas such as post-development, in which alternatives to development are proposed to achieve good living while coexisting with extractivist models [32]. This position is concerned not only with ecology but also with social justice, local autonomy and the questioning of capitalism and liberalism [17], but recognizes that in Latin America the scenario in which dependence on hydrocarbons and mining will be omitted and replaced by economies based on local alternatives such as agriculture or ecotourism is distant [15].

"Buen vivir" -Good living- is another recurrent quest in the global south. It is an alternative to Western-centric conceptions of development—all of them considered unsustainable—and to infinite growth [66]. It is about recovering ancestral practices

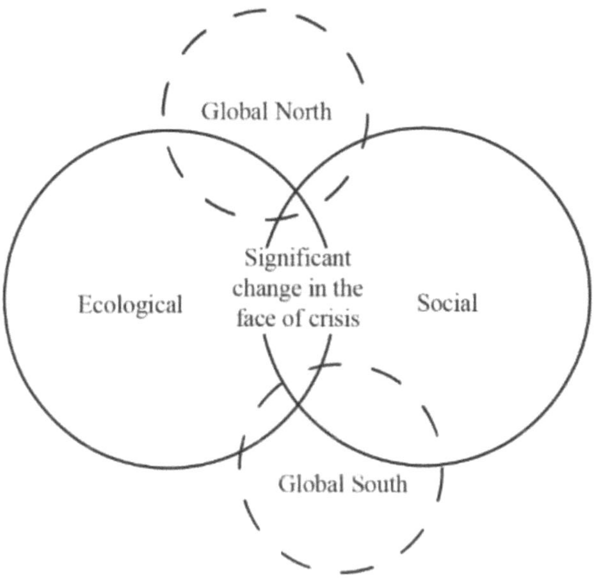

Fig. 2.5 Global territorial transitions. By the author, 2022

and knowledge of Indigenous communities and understanding them as contributions to a transition conscious of the finiteness of natural resources. This paradigm of southern transitional processes has also been addressed in the 2016 Peace Agreement in Colombia as one of the objectives for rural communities through a comprehensive reform and has also been included in the political constitutions of Ecuador and Bolivia. Gudynas and Acosta present it as an alternative to development and a "reaction against conventional developmentalism, as a bet on a substantial alternative" that separates from conventional Western ideas of progress including a special focus on Nature [32].

The design associated with this epistemological discussion includes practices that localize it so that it responds to unique and particular conditions of each socioeconomic environment as an alternative to universalizing visions. Practices such as ontological design, design for social innovation, social production of habitat and collective knowledge, together with their main contributions to the construction of the theoretical model for the purposes of this work, are summarized in Table 2.6 and discussed in the following paragraphs.

The understanding of differences must be part of the discussions for transition and of the design practices that contribute to it. Along these lines, the concept of ontological design presented by Anne Marie [81] includes the idea of a localized design in which the existence of different worlds composed of interconnected systems is recognized: the understanding of the pluriverse. Willis defines a "condition that is always situated [the condition of worldliness], and thus a starting point for understanding modes of human being such as dwelling and purposeful activity-for example, working

Table 2.6 Global South contributions to design for transition. By the author, 2022

Ontological Design	Pluriverse Ways of the human being Design of interaction Localized design	Collective Knowledge	Knowledge of the place Participation Education
Social production of habitat	Autonomy Collective ownership	Design for social Innovation	Co-Design: Expert + non expert Context based design Local economy

or designing" (p. 94) [81]. Arturo [16] for his part locates ontological design as "a means to think about, and contribute to, the transition from the hegemony of modern one-world ontology to a pluriverse of socio-natural configurations" [17].

One of the contributions presented by ontological design lies in the inclusion of the design of the interaction of the user or inhabitant with the space with assumptions such as: "The design with knowledge of the functioning of things will be qualitatively different from any type of design not known to it. Thus, one can move from ontological design as the naming of something to ontological design as practice" [81]. Placing ontological design as a practice and not only as a theory allows its implementation through different tools and devices that contribute efficiently to processes such as peacebuilding in Colombia. In this context, designs for the pluriverse become a tool to reimagine and reconstruct local worlds [17].

The questioning of the role of design in its various forms to strive for social, economic, and ecological innovation has also been addressed by design for social innovation. This design approach, unlike the previous one, focuses on the situation and addresses areas often excluded by the business and consumer market [36]. Likewise, it raises questions related to the future of human habitat design culture in order for it to be defined as truly sustainable. One of the main contributions of this approach lies in the location of an operational system in which expert and non-expert figures coexist and participate in a horizontal relationship. In this way, design for social innovation is assumed as "everything that expert design can do to activate, sustain and guide social change processes towards sustainability" [45]. This practice then places the designer as a facilitator and mediator rather than as an expert [17], promoting interdisciplinarity and the exchange of knowledge towards a common goal. Referring to this relationship between expert and non-expert in architecture, Schneekloth and Shibley argue that "even as we explore a movement beyond the culture of experts, we affirm the power of architecture to make substantial contributions to the messy vitality of everyday life in service of the promise of lives well-lived" [67].

In Latin America, the social crisis we are going through has also been reflected in housing deficit and precariousness. The housing models proposed by the state and large private construction companies have not contemplated the participation of the inhabitants, nor do they guarantee adequate housing. As a response to this phenomenon, a practice has emerged that promotes community participation, autonomy, self-management, and collective ownership called social production of habitat. "For millions of people, this is an alternative, sometimes the only one, to make access to a fundamental social right such as adequate housing a reality" [12]. The social production of habitat is, consequently, "one of the disciplines called to the formulation of applicable tools that allow overcoming the physical and social inequality of urban territory, resulting from its entire competence" [68].

Finally, in Colombia García-Reyes and Anzellini proposed a design practice that seeks the joint construction of knowledge through architectural and technical solutions consistent with the context in which they are located. *Saberes Colectivos* is methodologically based on the active interaction of inhabitants, technicians, and professionals, to achieve an understanding of the physical, spatial, and environmental conditions of the place and the practices of inhabiting. This interaction occurs at all stages of the process: recognition, projection, construction, and legitimation [27].

This methodology integrates the local issue as a fundamental basis for project success. Listening to and understanding the needs of the inhabitants, and placing them in horizontal dialogues, can generate appropriation by the beneficiaries of the projects, reducing their risk of abandonment in intermediate phases or the perception of non-ownership of the built spaces by entities outside the territory [15].

Some of the concepts present in these approaches and practices find antecedents in critics of modernity such as Bernard Rudofsky and his work for the New York MoMA exhibition *Architecture without Architects, a brief introduction to architecture without pedigree*. There, Rudofsky "develops an appraisal of the vernacular as an alternative to the Modern, questioning the contribution that the architect as author has in shaping our built environment" [43]. Rudofsky highlights the need to value traditional design practices in the absence of the architect making a conceptual contribution to the construction of vernacular architecture and dares to confront modernity having surpassed the first half of the twentieth century.

Similarly, Colombia already had precedents such as Orlando Fals Borda, who promoted and applied IAP: Participatory Action Research, a research methodology that profoundly questions the traditional approach to the subject of study—community, people or individuals—with airs of interventionism and/or scientific colonization to be reinterpreted with tools such as collective research, critical recovery of history, the valuation and application of popular culture, and the production and dissemination of new knowledge [20]. Priority would then be given to group discussion, collective memory, oral, cultural, and musical traditions, mutual respect, and popular knowledge as reliable sources of information and as tools for social construction exchange.

Likewise, Giancarlo [13] also exposed a position on participatory architecture, an emerging concept in the second half of the twentieth century, stating that contemporary architects "must do everything possible to make architecture less and less the representation of its designers and more and more the representation of its users" [13]. This idea gives a political sense to the practice of architectural design as far as it implies the recognition of public interests and the search for spaces that represent them.

Although the practice of design in recent decades has included the discussions presented so far, the speed with which crises arise demands a permanent innovation in the same design practices. Likewise, it is necessary to adapt conceptual and methodological approaches to specific contexts such as overcoming a rural armed conflict. This also requires innovation in the tools and devices used in the design exercise so that their contribution to the transition is greater.

The lack of guarantees in participation is not, however, an exclusive problem of the Global South. Although to a different extent, the Global North also faces what Jeremy [70] calls "appeasement participation," which consists of the mere intention to make communities feel involved in urban and/or architectural development decision-making processes through assemblies or consensuses that, in reality, have no impact and are mere formalities behind economic or political machinery with decisions already made [70]. This "pseudo-participation" does not have a valid counterpart in the North, since, in his opinion, "architecture will always have individual interests, inherently variable politics, and a problem of relating and communicating assertively between expert and non-expert" [70].

From another perspective, in the Global North, there are proposals such as the (Re)production of Architecture by the academic Doina Petrescu, who understands our current era as a time of crisis in which not only the economy, the ecosystem, and politics are impacted, but also our current system of production and reproduction. In that sense, she argues, the transition is not just an option but a necessity, which, although still diffuse, must find ways to be sustainable for our survival, and that is where architecture must be reinvented and rethought in practice [58].

These variables are considered in the proposal to think of architecture as a system that is, according to Angelika Fit (2019), under "intensive care," and that should also adopt care policies that arise from the individual's own perception of their individual development to the impact they have on the conservation of the habitat, the species, and the way we live under the premise of mutual care [22].

Although these new approaches to architecture pose a strong critique of traditional modes of design, participation, and dominant scientific currents, "it is not about discrediting science or an essentialist anti-science fundamentalism" [65], but about understanding and implementing what the author coins as the Ecology of Knowledges, a scenario where it can dialogue with "lay knowledge, popular knowledge, indigenous knowledge, the knowledge of marginalized urban populations, and peasant knowledge" [65].

Today, the legitimation of design practices for transition in the architecture of the global south is evidenced in the sense of appropriation by the user and the community that will inhabit the designed space. To that extent, architecture can truly contribute to a social transformation based on autonomy and vision for the future. The winner of the Pritzker Architecture Prize for the year 2022, Diebedo Francis Kéré is one of the most outstanding recent examples of the potential that architecture has when it is based on community design practices and arises from local knowledge and traditions. His work is based on localizing design by recognizing the physical and cultural conditions of the environment and innovating in techniques and technologies that seek the permanence of space and activity into the future. “Design needs to respond to the current situation and, at the same time, anticipate future growth” [39].

Kéré also integrates in his practice the relationship between the expert and the diffuse designer proposed by Manzini earlier, but further recognizes that architecture can be learned by anyone, and that architecture can make a difference in people’s lives if it is grounded in the local climate and culture [39] Architecture of this type “can activate processes of change that serve to break with hegemonic Western models, which are often adopted in an uncritical and mechanical way and do not respond to the reality of developing countries” [11]. In an interview given in 2017 Kéré further proposes bridges between the architecture of Burkina Faso and Latin America, stating that there is a common urgency to improve the quality and accessibility of housing, as well as “there is a strong will to take care of our people. There is an awareness and pride in our cultural heritage, we recognize it as something to be celebrated and preserved for future generations. Architecture can help achieve this” [39].

Community appropriation is then configured as an engine that drives restorative practices in complex contexts. The Casa del Pueblo in El Salado, Colombia, is a notable example of this. In 2000, a massacre took place on a sports field, the only paved area in the town, killing more than 100 people and forcing the displacement of 300 others. The community agreed that no monument could replace or commemorate what had happened. However, one act managed to unite them: the cleaning of the field. “Only after the cleaning of the court the community felt able to continue building actions” [4].

In 2012, accompanied by architect Simon Hosie, they decided to transform the meaning of the place by changing the name of the place where the massacre happened calling it Campo de la memoria (Field of Memory). Subsequently they proposed the Field of Culture around a huge millenary tree. In this field they built La Casa del Pueblo, which functions as a library and where even the furniture was designed by and for the children, who are the users.

For this intervention, Hosie used a participatory design tool called *Planos Vivos* -living plans-, which combines technical, urban, and historical information of the town, as well as anthropological aspects such as the meaning of spaces, landscape elements, materials and notions associated with customs and behaviors. “The meaning given to the place propitiates solidarity among the surviving victims of the massacre and evokes the presence of the loved ones taken from them” [34].

This is one of the tools that will be presented in greater depth in chapter four, selected precisely for the verified or potential contribution it provides in the generation of sense and meaning to spaces to strengthen design practices in the transitional context.

The emergence of these and other emerging practices account for the awakening of designers in the face of the ruptures inherent to transitions. The peace process in Colombia, as a transitional process that aims at the transformation of Colombian society and the construction of a lasting peace, should be seen as a scenario of action where design is also concerned with contributing to overcoming the conflict. Design for transitions implies a transition of design towards one that is based on local and localized practices centered on the user; a design that promotes the autonomy of communities from the state; that emerges in a collaborative manner in a horizontal system facilitated by experts; a design that generates dialogue of knowledge and construction of knowledge; a design that identifies the particular needs of the different groups and communities involved and includes them in the projects development. Within the framework of the implementation of the Peace Agreement in Colombia, architectural design has the responsibility to contribute to the non-repetition of the conflict.

2.5 Architecture and Peace Agreements. International References

As mentioned in previous sections, social inequality, lack of access to land and land tenure, and economic and political interests have been the main causes of internal armed conflicts in the world. These are mainly located in regions where this socio-political scenario is recurrent, such as Africa and Latin America. Likewise, the negotiated solution to some of these conflicts has made it possible to listen to the voices of the actors directly and indirectly involved and to agree on actions that favor spatial interventions such as the improvement of habitat, infrastructure, or equipment. In this section, we present a review of two Latin American cases, emblematic for the way in which the conflict is resolved through the signing of an agreement that includes the generation of different architectural projects: Guatemala and El Salvador. The review will allow us to gather some experiences that can feed the approach of the theoretical model that is the objective of this chapter.

These achievements underscore the need to understand peace not only as the absence of armed conflict but as the convergence of the guarantee of human rights, social justice, and the "creation of conditions that reduce structural violence, inequality, and injustice" [62]. From this perspective, innovative and holistic research on conflict is necessary to recognize all its facets, including not only the diplomatic but also making use of fields such as anthropology, political science, and economics. Likewise, it is necessary to evaluate dimensions of the scale and scope of the conflict, overcoming the "traditional distinction between internal and external conflict" [62].

2.5.1 Guatemala

Peacebuilding in Latin America can be seen as a relatively new practice. In this first case, it was on December 29th of 1996 that the government of Guatemala pursued the end of the armed conflict with the Guatemalan National Revolutionary Unit, known by its acronym URNG, a leftist armed group born in 1982 conformed by four already existing guerrillas which altogether, had been operative for 33 years by then.

The goal of the Agreement for a Firm and Lasting Peace, was to "put an end to the violence which had taken more than 200.000 lives" [83] which according to the URNG, "was never part of the purpose or the strategy to fight exploitation, race discrimination and cultural and political oppression" [75] allegedly attributed to the state and amongst other reasons, the political intervention of the United States in the country in 1954 which was perceived as a "brutal repression and ideological control" [83]. Born on these premises and lead by the agrarian working class and Indigenous people, the civil war lasted more than four decades until in, 1987 under a conjoined group of Central American countries seeking peace: Costa Rica, El Salvador, Guatemala, Honduras, and Nicaragua, the Esquipulas II agreement, opened a window of dialogue within each participant country.

The creation of the Reconciliation National Committee -CNR-, was the opportunity for the first contact of the two parties in Madrid in 1987 and the beginning of a 9-year long process with the intervention of international folk countries and UN participants as seers. Within the process itself, members of the URNG and the Guatemalan state, met with different sectors of society. These dialogues with the catholic church, unions and popular organizations, members of academia and small businesses, laid the foundations to reach a peaceful negotiation under the general concerns of polarization, due to multiple actors involved and, in some cases, absent.

In these early stages, the process was decisive with the consensus of 3 initial points: global agreement on human rights, relocation of displaced communities and establishment of the Commission for the Truth. The second point in particular, comes as outstanding since it not only meant the one addressing directly to the territory as a part of the conflict, but as a peacebuilding scenario with "pledges for the uprooted people promoting their productive integration and the creation of a technical commission with its funding for the execution of the agreement" [51].

Nevertheless, under the circumstances of the state at the time, these goals were seen as too ambitious and somehow void in form on later post agreement stages. Many agrarian organizations expressed their frustration on the demand for a thorough reform in land access because as the nature of these reforms "was only possible under a liberal and representative democracy but not under a model that questions neoliberalism or the pattern of accumulation", change didn't seem as a realistic goal [82].

As the process went on, it got more evident how Guatemalan democracy was tainted with two obstacles, institutional barriers, and ideological barriers [51]. The first, addressing the political system which denied minorities in participation such as the indigenous peoples and women, and the access to equal conditions,and the latter, prejudices, and stereotypes on the social scope inside one of the most multiethnic and multicultural countries of Latin America where around 43% of its population belong to an indigenous group. It became undeniable that for the purpose of a "positive peace", it was urgent to switch to a new collective mentality not only for the sake of ex-combatants re incorporation but for a sectorized country [46, 82, 51]. On this point, the issue of housing and land ownership was in fact a priority agreed upon in the accords, which established "special attention to guaranteeing access to housing for demobilized combatants who enter rural areas and to granting adequate credit facilities to those who enter the urban environment" [26]

With this under consideration, the Program for the Community Development of Peace -DECOPAZ- was born with the financial and technical assistance of the Inter-American Development Bank -IDB- whose experience on international similar cases, worked as lessons to be implemented in the search of development, management and execution of social projects and housing. The goal was the "physical, human and social recapitalization of the most affected areas by the armed conflict" [47].

It's worth noting that not only the funds would be put under public management and that people would be educated for the effective execution of these projects, but also, how there would be "a participative planification to promote a culture of collaborative work in a peaceful environment" [47]. This was indeed unprecedented and as it was to be expected, not easily embraced by a wounded community which lacked trust on governmental actions and false expectations created by institutions with similar promises sometimes with criminal intentions in the past. With this harsh context ahead, multiple NGOs were hired to follow up on this process for several years and create a trusting environment with training and technical assistance on communities.

The strategy was to focalize attention into trusted community leaders to make them enroll and take participation. Then, as a spatial tool, the program developed a game plan for small portions of the territory called Micro—regions which facilitated the focus on specific demands and sensibilization of the micro regional planning along the community to prioritize urgent spatial needs.

After several years of empowering communities through community diagnosis, self-governance, management of public funds, investment plans, training and collaborative planning, the results were not as good as expected as by the year 2000, which was the stipulated date for the conclusion of the program, less than 50% of the projects were inaugurated. This was attributed to the following reasons mainly:

- The programmed schedule for the activities for each phase of work was not coherent with the conditions of the community [47].
- Lack of financial responsibility by DECOPAZ which at some points stopped the payment for ongoing construction projects.
- Not a comprehensive view of what a transparent community process meant: Confusion on specific tasks and ambiguity, limited and slowed down the participation of the community.
- Uncommitted communities which failed to coordinate for operative tasks such as the storage, transportation, and general care for material.
- Unrealistic execution deadlines, since a very ambitious timetable was set for implementation.

On a brighter side, it was plausible how the post agreement era, activated social movements which managed to resist and strategize their participation, this time, "with public laws, social communication and political education so much needed to increase the level of political awareness amidst the working classes" [82] especially on the land access issue.

This positive change in participation echoed and persisted in time as people in distant rural regions became active in decision making through referendums. For instance, "between 2005 and 2013, 73 municipal and community referendums were successfully carried out with an absolute majority of people rejecting projects of (oil) exploration and exploitation of natural resources" [82]. These kinds of ideals put into perspective the reborn need of not only claiming back the territory taken during the armed conflict, but its defense and appropriation within a healthy natural environment for the benefit of its inhabitants: a real promotion of autonomy.

Other successful cases come into consideration such as the North–American Hug it Forward NGO which through a simple artifact: plastic bottles filled with inorganic trash, has managed to build schools in marginalized areas of the country. Not only it boosts environmental consciousness on correct disposal of waste but uses architecture as a means for participation as the whole community takes part in construction.

Under this general picture, Guatemala's case study shows very particular lessons on participation, self-governance and community engaging. And so, it shows "the limited institutional and legal modifications of the post war years was the result of the efforts and fight of social movements, but it did not change the historically rooted structures of power" [82].

2.5.2 El Salvador

El Salvador's peace agreement was one of the most politically relevant changes in Central America. The socio-political scene of the country had been under a military oriented regime since the 30's. Its violent and oppressive ways had led to a context of fear and intimidation to avoid uprising from the marginated sectors of society such as the agricultural and working class. Every single attempt of democracy was soon turned down by what was starting to become an oligarchy in the hands of the richer. The state itself, was a promoter of exclusion, inequalities, and violation of basic human rights.

Under such circumstances, a leftist movement was born and quickly adopted a guerrilla pattern: the Farabundo Martí National Liberation Front—FMLN, named after a communist and revolutionary politician active during the 20's. From its conception in 1980, the group stated its upfront "final offensive" against the Salvadoran government which of course, did nothing but increase military actions escalating the existing conflict [42].

Operating from the rural areas and mountains of the country and gathering militants from Marxist-Leninist ideas sympathizers, its power increased to the point of becoming an actual structured army by 1985, with enough power to stand as a true opposition for the government's regime. Its idealistic roots started to fade as the main goal was to take power by force which prioritized armed actions instead of political ways. At least 75.000 people were killed during the civil war, another 15.000 reported as missing and around 500.000 forced to leave the country. This sets an alarming number of approximately 2% of the total population of El Salvador as a victim of the armed conflict, by its total population at the time.

During this discouraging panorama, various attempts for dialogue were made from international organizations, the catholic church, and the new constitution of 1983 as a basis for change in the country's political landscape opening a gap for opportunities. With these statements at hand, the president of El Salvador, Jose Napoleon Duarte, called out to a peace dialogue in 1984 which was accepted the next day by the FMLN, creating the first scenario for an encounter in the middle of armed conflict in La Palma municipality.

The very first stages of the dialogues did not come up as productive. For the first five years, there was not an agenda to follow but simply meetings among the leaders of both sides with no actual or palpable results. The violence was still ongoing and in 1989, the biggest direct confrontation took place in what was called the Final Offensive, which was a violent outbreak of almost a month with a series of assaults from FLMN to the government. It was, nevertheless, a trigger for the second and most productive stage of the agreements. With an actual agenda with specific topics, deadlines and international watchers, the dialogues continued until 1992 when they were finally signed upon.

In spite of the media coverage of the agreements to public knowledge, the negotiations were not an opportunity for a social dialogue and the agenda didn't contain citizens' concerns or representatives per se. Ending the armed conflict was always the priority and only after the final agreement, economic and social aspects were brought to the conversation with specific programs and forums. One of them, the National Reconstruction Plan -PRN-, directly addressed the issues of land, infrastructure, and basic needs coverage. Despite its "dependency on international funding and cooperation, which would run scarce with time" [79] there would be major accomplishments and successful projects.

Taken from an economic perspective, the government focused on creating stimulus and funds for construction companies to facilitate social housing credits like the National fund for social housing -FSV-. The objective was to delegate the lack of housing to private companies which would, in the 90's boost, a rapid and intense construction boom in the periphery of San Salvador. One of the to quickly catch up with these incentives, was the construction conglomerate "Grupo Roble", which with the help of the government's new laws, was able to build, plan and execute around 5 master plans for housing in metropolitan San Salvador initially and then 8 others for a total of around 65.000 units after the peace agreements.

With the aim of a diversified offer of housing typologies and prices, social housing purchase has had a significant impact on the growth of the city and the dignification of living conditions of many low-income families and victims of the conflict. It is also outstanding, how urban policies and laws were reformed in order to gain access to a fair and new type of land distribution and acquisition by privates.

More recently, efforts were made to increase autonomy on the construction or improvement of housing through policies and funds like the Veterans and Ex-Combatants Fund signed by the current president and given by the National Fund for

Popular Housing -FONAVIPO- in early 2020 under the Ministry of Governability and Territorial Development.

These efforts both by private individuals and the government, are an example of purely economic aid. Nevertheless, international NGOs such as Habitat for Humanity, have been present when the alarming rates of a decent housing conditions has increased over the years and now 39.2% of families in El Salvador, live in overcrowded homes [33]. By empowering and training women and local communities, the projects have been able to improve living conditions and raise awareness for the urgent housing crisis.

This particular case study shows a different approach to the role of infrastructure and architecture in the post conflict scenario where political and monetary reforms are the base for sustaining a habitat for peace building. With a lack of community engagement, the projects might not be suitable to a real placemaking or self-governance context as no participation in design or construction has been a part of the solution. Nevertheless, over the initial years of its implementation, the goal was met and both victims and victimizers from all socioeconomic rates had access to a dignifying housing solution.

2.6 Theoretical Model Consolidation

Modern and contemporary architectural design paradigms still assume architecture as a formal, spatial, functional, and technical problem. However, in a scenario of global transition characterized by social and ecological crisis, architecture must assume a transformative attitude through projects that respond to local conditions and needs in direct and deep connection with the territories and their inhabitants. In this sense, architectural design for transition emerges as a possibility for thinking, proposing, and executing territorial transformations from a broad and holistic understanding of concrete contextual realities.

In addition, architectural design in a framework of transition towards peace such as the one Colombia is experiencing appears as a practice of enormous responsibility in the contributions to the construction of sustainable ways of living. These must be focused on the non-repetition of the conflict, minimizing the risks of relapsing into the socio-political and physical-spatial context described in the first section of this chapter, which favored the consolidation of the war in rural areas. In a transition process as particular as the Colombian one, architectural design must consider innovative criteria and variables in line with local and community demands that promote good living, autonomy, human solidarity, social equity, affinity with nature, among others. This section aims to bring together the discussions developed so far and propose a theoretical model that serves as a conceptual representation for the Analysis of architectural projects generated in rural areas of conflict and post-agreement in Colombia and the subsequent approach of architectural design tools for this context.

2.6.1 Conventional Architectural Design Variables

Architectural design in a transitional context of peace building in a country like Colombia requires the revision of the conventional variables of architecture positioned in the aesthetic and functional search for the project. From there, it is important to establish correlations that bring the project closer to a social and political approach to improve the scope and contribution of architecture in the common purpose of overcoming the conflict. "By persevering with a poetic approach, architecture has become less political. So, when it addresses the social, it does so with education. Social change will not come from a strictly artistic approach" [31]. These variables are associated with paradigms already installed in architecture that allow us to characterize design in terms of form, space, function, and technique. By means of these variables it is possible to measure, qualitatively or quantitatively, the minimum verifiable aspects of any architectural project. However, in the construction of this theoretical model, these variables are fed with some of the parameters of design for transition identified in the previous sections and located in the context of the conflict and post-agreement in Colombia.

Variable a—Form and Space

Form and space have been extensively studied in the history of architecture from the different aesthetic currents. From antiquity to the present day, they are obligatory subjects of the great theoretical treaties. In this sense, we can affirm that form and space are inherent to architecture. In other words, form and space are the "basic means of architecture that include an elementary and timeless design vocabulary at the same time" [10]. However, understanding formal and spatial needs in a post-agreement scenario implies understanding in a particular way the dimensions that configure them.

This variable is composed of dimensions such as architectural typology, spatial and volumetric proportion, enclosures, principles of order, implantation, among others. These dimensions can be measured qualitatively or quantitatively, depending on the case, with indicators such as distribution forms, height, or area. These characteristics of rural architecture in Colombia have been linked to the customs and ways of life that condition the way in which the form of spaces is defined, almost always without architects. In these areas, the scene of armed conflict for decades, housing, community facilities and productive spaces were mostly built without aesthetic pretensions, but with empirical responses based on local knowledge. In some institutional buildings such as schools or hospitals, form and space are also conditioned by normative aspects exercised by the institutional framework, and little considered in rural domestic architecture.

Faced with the challenges of design for the transition to peace, form and space in architecture must consider dimensions that respond to the need to adapt to the different conditions of the environment. Thus, the adaptability and flexibility of spaces can allow adaptations for different economic activities and productive projects in the implementation of points one and four of the peace agreement, related respectively to the Comprehensive Rural Reform and the Solution to the problem of illicit drugs. The design of architectural spaces with progressive growth allows the inhabitants of a community to design and manage their vision for the future individually and collectively. Table 2.7 shows the synthesis of this variable.

Variable B—Activity and Relations

Architecture arises in a primitive way as a response to a need for protection and shelter or to the conditions for carrying out a specific activity. This paradigm has been reinforced in history with rationalist principles such as *form follows function*, placing activity and relationships as the initial variable in the architectural design phase. In that sense, we can say that the architectural fact requires the existence of an activity or need to be conceived. Some of these needs are objective, such as social, cultural, biological and site needs. However, architecture must also respond to subjective needs that are the result of the freedom, autonomy, and peculiarity of each individual [41].

In this variable, dimensions such as interaction with space, the architectural schedule and organization chart, and accessibility and connections are proposed. Some of the indicators used to measure it are the percentage of circulation areas, the area/person ratio, the number of accesses, the number of spaces, the number, and types of uses, among others. With respect to this variable, domestic and communal spaces in conflict and post-agreement zones have emerged with an initial functional purpose which, in some cases, was affected by the dynamics of the war itself, such as forced displacement, dispossession of property and illegal economies that required the adaptation of spaces designed for another use. However, the function of space in this context has been linked not only to basic housing needs but also to productive rural activities such as agriculture and livestock raising, community activities such as meetings and recreational and leisure activities, even during the armed conflict.

In this sense, this variable must understand the differences of each person and each place in order to respond to the ways of living of the inhabitants and their specific needs in the transition from war to peace. Multifunctionality is therefore a useful indicator to measure the way in which architecture contributes to points one and four of the Peace Agreement in the search for the non-repetition of war economies through useful and adequate spaces for the consolidation of emerging productive

Table 2.7 Variable A—form and space. By the author, 2022

	Parameters	Indicators
Variable A: form and space	1. Flexibility and progressiveness of space 2. Architectonic typology 3. Spatial and volumetric proportion 4. Ordering principles 5. Enclosures and openings	1. Possible use of space 2. Expansion areas 3. Ways of distribution 4. Height and number of stories 5. Built areas

projects. Likewise, the projects must include consensual decisions in relation to the architectural schedule and the activities associated with the ways of living in each place. This understanding can be seen in the appropriation of space through modifications and adaptations in the activities in line with the requirements of each time and place. Table 2.8 shows the synthesis of this variable.

Variable C—Technique and Technology

Technique and technology in architecture comprise all the processes and materials necessary to build a work. To this extent, there is no architectural work without the appearance of this project variable. It can identify dimensions such as comfort, the construction and structural system, the materiality and feasibility of the project, among others. The indicators include values such as thermal insulation, structural strength and stiffness, the amount of material used and costs.

Territories in transition should be scenarios for the continuous generation of techniques and technologies that seek to reduce the ecological footprint of construction. Although this is not a recent or exclusive search for the peacebuilding process, it is relevant for the purposes of this theoretical model to demonstrate the use of eco-technologies as an indicator for measuring projects, with the inclusion of processes such as the use of rainwater, solar energy and waste, among others, given that in rural Colombia there is a common deficit and precariousness of basic public services such as water, sewage and electricity, and architectural projects must meet these needs. An example of these practices could be seen in the review of the case of Guatemala and the participatory experience of building with plastic bottles filled with inorganic waste. The innovative presence of responses to these dimensions aligned with environmental protection in the country is also one of the purposes of the Comprehensive Rural Reform included in the Peace Agreement in Colombia.

In addition, in the specific contexts of peacebuilding, it is important to contribute to the appropriation of spaces not only in terms of function but also in terms of processes. In this sense, the presence of local construction practices and local labor is fundamental, recognizing the different knowledge of the inhabitants. To the extent that the user is part of the processes of management, design, and construction of spaces, it can be expected that they feel it as their own, take care of it and use it. Respecting and valuing the local environment also includes the use of vernacular

Table 2.8 Variable B—activity and relations. By the author, 2022

	Parameters	Indicators
Variable B: activity and relations	1. Understanding differences and ways of living 2. Consensual decision making 3. Interaction with space 4. Architectural program 5. Architectural schedule 6. Accessibility and connections	1. Capacity of multiple uses 2. Appropriation and adequacy of spaces 3. Percentage of areas of circulation 4. Quantity of spaces, accesses, and (types of) uses 5. Area/person ratio

Table 2.9 Variable C—technique and technology. By the author, 2022

	Parameters	Indicators
Variable C: Technique and technology	1. Ecotechnics 2. Acknowledgment and appreciation of the local 3. Comfort 4. Constructive system 5. Materiality 6. Project feasibility	1. Taking advantage of rainwater, solar energy, reusable waste and local materials 2. Taking advantage of local practices, vernacular techniques and knowledge 3. Thermic insulation, resistance and structural rigidity 4. Quantity of used material and costs

techniques to strengthen construction cultures and the reciprocal training of all the actors involved in the development of a project through the dialogue of knowledge. Table 2.9 summarizes this variable.

2.6.2 *Architectural Design Variables for Transition*

Architectural design for the transition to peace in Colombia must consider unique aspects of the contexts of action in order to propose localized solutions that respond to the particular needs of communities. Form, space, function, and technique can be standardized and materialized in replicable typologies subject to transformation. However, every context is different, every community is different, and every person is different. In this order of ideas, it is required that the design considers variables rooted in the ways of being and inhabiting the territory. Sense and meaning are proposed as useful variables for this part of the theoretical model as they are not standardizable and complement the conventional variables of architecture to contribute to the generation of appropriation of space, the vision of the future and community autonomy.

Sense and meaning

Integrating discussions of sense and meaning into architectural design implies recognizing that architecture is not only perceived aesthetically but also semiotically and politically. Design in a transitional context must therefore understand that architecture responds to local needs and represents local cultures as well as their construction practices, their historical memory and their vision of the future. Contributions such as those gathered from the experience of El Salvador allow us to demonstrate the importance of the generation of sense and meaning from below, transcending the mass and industrialized production of architecture that is perceived as transitory and does not generate roots. To integrate these variables into the model, it is also necessary to synthesize the way in which they should be understood in the context of the transition to peace in Colombia.

Meaning refers to a conceptual representation of something, while sense corresponds to the intention or interpretation given to that representation. According to Roland [6] signification, which gives meaning, corresponds to "the obvious", that which is intentional, and which exists in function of the addressee of the message, while significance corresponds to "the obtuse", because of the relation with the angle which opens the field of sense infinitely [6]. Along these lines, a school will have a common meaning referring to the definition of space, but for each student and for each teacher in the school, that space will have a different sense. The architecture built in the implementation of the peace agreement, for example the ETCRs, signify an instrument of action for reincorporation, but this instrument must be interpreted differently as for each inhabitant it has a different sense to be there.

The integration of sense and meaning into a single variable is proposed to allow for the qualitative evaluation of parameters and indicators. This is because sense is the intention and interpretation that the user will give to the space, the meaning of which must be constructed in the design practices. Sense is given by an intimate relationship based on lived experiences in the space. For example, a community center whose meaning is related to a meeting place, changes its sense when in that same space meetings are intended to intimidate the inhabitants by different armed actors during the conflict. Or a sports field where a massacre took place. Likewise, in peacebuilding, a place that was the scene of a guerrilla attack in the past can become a community vegetable garden, an ETCR or a memorial. A place is thus determined not only by physical-spatial limits but also based on a vital experience, where relationships between individuals generate ways of acting, living, thinking, feeling, and knowing, and other dimensions of individual and collective everyday life [cultural, economic, political, etc.] are considered [28].

In order to measure and evaluate these variables in design practices for the transition to peace, dimensions such as the promotion of autonomy, co-design and participation, training and dialogue of knowledge, vision of the future, appropriation of space and localization of design must be considered, which will be described and justified in the following paragraphs.

Promoting autonomy. To the extent that each community promotes the development of its own resources and tools to take the next step, it will be able to advance towards autonomy in the processes. Creating the conditions for self-management and maintenance of infrastructure, equipment and housing can contribute to future sustainability. Likewise, the community must lead these processes based on a long-term vision. To this extent, the design of the interaction between the user and the architectural project proposed by ontological design must also be considered, which proposes that "design with knowledge of how things work will be qualitatively different from any kind of design that does not know how things work" [81].

In the peacebuilding process, achieving autonomy does not mean disconnecting from state centralism, but rather reducing dependence on the state to minimize the impact of the marginalization that has favored the emergence and consolidation of the armed conflict. Likewise, achieving self-management without an adequate localization of problems and solutions can minimize the impact of projects, as can be seen in the case of Guatemala, where the community achieved autonomy in some processes, but the results were not what was expected.

Co-design and participation. In architectural design for transitions, expert design interacts with non-expert or diffuse design in a horizontal relationship in which the designer sheds his or her particular interests and positions him or herself as a facilitator of processes guided by the interests of local inhabitants. This allows for an interconnection and an integral development of the projects, favoring the appropriation of the processes by the stakeholders. This requires more open and collaborative processes that can create opportunities for democratic action and celebration of everyday life [67]. "Design involves meeting material needs and resolving competing social requirements through a process of active participation of occupants and the mediation of "professionals" who accept responsibility for the inevitable production of identities, communities and cities through architecture" [8].

Collaborative and participatory practices in architectural design for the transition to peace in Colombia can be evidenced not only with ambitious large-scale and long-term projects but also with ephemeral or small-scale alternatives that address universal and mass solutions [4]. These alternatives can foster mutual respect by strengthening relationships around local knowledge. Planning with people makes architecture resilient to the adverse circumstances and changing times. [4] and that people take control of the events and circumstances that take place in their lives [67].

Architecture designed and built with the community, not for the community, will facilitate the inclusion and resolution of local needs and reduce vulnerability to conflict. The empathic relationships present in co-creation actions can generate an effective reconciliation mechanism for communities in conflict [57]. Likewise, understanding architecture as a device for local political action rather than as an aesthetic pursuit will allow for collective growth in local autonomy and decision-making.

Training and exchange of knowledge. To the extent that design practices take place in a collaborative manner, a dialogue between the knowledge of the people involved is organically produced. These practices can also be considered as objectives of architectural projects within this framework, including pedagogical strategies that promote the reciprocal training of inhabitants and experts. In this way, the sense of the project is broadened, and capacities are strengthened that will enable the self-management of future projects. In the processes of reincorporation and peacebuilding, people who developed different skills during the war years, inside or outside the armed organizations, are present. This opportunity was taken advantage of in the case of Guatemala, for example, where education and participatory planning were key to the implementation of projects.

This creates unique opportunities for reconciliation, reincorporation, and the generation of spaces for continuous collective learning. In this way, space emerges as a social product configured through the superimposition of discourses and practices, configuring space as a constituent part of collective life and not only as a static framework for it [28].

Vision for the future. In the context of building stable and lasting peace in Colombia, projects must integrate solutions designed for the non-repetition of conflict. This is because design solutions for transition are rooted in long-term thinking, are lifestyle-oriented and place-based, and always recognize the natural world as the broad context for all design solutions [36]. Points three and five of the peace agreement, which speak respectively of the end of the conflict and the victims of the conflict, can receive significant contributions from architecture that responds to this parameter mainly from the understanding of memory and local capacities and the inclusion of sustainable proposals in time, progressive and localized. The architectural project must become a way to tell the stories, facts and narratives that shape and nourish the space and the collective imaginary [5]. Likewise, architecture must "build the absence of the acts that will remain in the past, of those who lived and suffered the conflict, of the victims. To build memory for those who will remain, a place for that absence to be evoked and remembered" [37].

Undoubtedly, architecture for memory has an indispensable role in non-repetition, but architectural design must also be located as a practice that transforms the autonomous processes of communities that build from and not only for memory. There, design becomes an everyday practice that generates a vision of the future, rootedness and appropriation of the territory and the ways of inhabiting it again, in the absence of armed conflict. Place-based cultural, ecological, and economic practices are important sources of possible visions and strategies for the reconstruction of local and regional worlds [16].

Appropriation of space. Architectural projects are not finished when they are built. On the contrary, it is when the space is materialized that the life of the building begins, starting with the interaction with the user. In the peace-building scenario, the projects that are executed must be felt as their own by the community from the moment of their management and design so that they are also felt as their own afterwards. To the extent that there is ownership, the space is susceptible to permanent and consensual modifications and adjustments to adapt to the dynamic needs of the territories. Buildings constructed by external experts that do not achieve this link run the risk of not being received as part of the user's everyday life.

Participation and collaboration can then ensure ownership of space. "When we plan for people we tend, once consensus is reached, to freeze it into a permanent fact. It remains rigid no matter how liberal the initial intentions are. But if we plan with the people, the consensus remains permanently open; the act becomes liberating and democratic, stimulating a multiple and continuous participation; a political action" [67]. Communities must participate and nurture their development in order for architecture and public space to take root. [4] and have a sense of belonging to the territory. In this way they will not abandon it so easily, they will defend and nurture the processes that strengthen it and will be able to follow their vision for the future.

Localization of the design. Unlike the localization of the built object, it requires an understanding of determinants and conditioning factors that are localized in a place and belong only to it, that cannot be standardized. This implies contemplating the future vision of the users, the capacity to generate autonomy, local knowledge, ways of inhabiting, in short, an understanding of the pluriverse. In this quest, architectural design must promote permanent participation in all phases to facilitate the consideration of these dimensions in the architectural project.

Acknowledging the real complexities and contradictions inherent in a place continually negotiates meaning and position, including where we place ourselves as experts [67]. Likewise, a localized design practice remains, evolves and is autonomously managed once the expert withdraws from the process. Localized design necessarily requires dynamics that emerge from within but must always allow for the bringing together of prior knowledge and experience in the service of collaborative construction. In this way, the ordering of the world proceeds according to a local-to-global logic [54].

Localizing design also implies decolonial thinking in which centralist and western practices of producing space are questioned. In such thinking, "subjects, collectives and social groups living on the margins produce knowledge that can be transferred from one place to another, constituting an exteriority to modernity" [14]. This thinking also favors the recognition of differences and questions relations of domination, inequalities, and socio-spatial conflicts in order to transform them [14]. In this way, place refers to a broader social context, the local dynamics, assigning a relevant sense to territory and culture as political elements to challenge, associate, or distance themselves from the hegemonic discourses of modernity [28]. Table 2.10 summarizes this new variable.

Table 2.11 summarizes the proposed theoretical model and the interaction between the conventional variables of architectural design, the variables of design for transition and the dimensions and indicators designed to measure them. The model can be used not only for the evaluation of projects executed in the framework of the implementation of the Peace Agreement, but also as a dynamic tool of architectural design for transition in peace-building processes such as the one Colombia is initiating, particularly in territories where the discussion of alternatives to state neglect, marginality and different territorial conflicts is present.

Table 2.10 Variable D—sense and meaning. By the author, 2022

	Parameters	Indicators
Variable D: sense and meaning	1. Encouraging autonomy 2. Co design participation 3. Education and local knowledge dialogue 4. Vision of future 5. Appropriation of space 6. Location of design	1. Self-governance methods 2. Quantity of consensual decisions 3. Implemented pedagogical practices 4. Interaction of user space 5. Collaborative design practices (quantity, depth, complexity, continuity, etc.)

Table 2.11 Theoretical model of architectural design for transitions. By the author, 2022

Variable	Parameters	Indicators
Variable A: form and space	1. Flexibility and progressiveness of space 2. Architectonic typology 3. Spatial and volumetric proportion 4. Ordering principles 5. Enclosures and openings	1. Possible use of space 2. Expansion areas 3. Ways of distribution 4. Height and number of stories 5. Built areas
Variable B: activity and relations	1. Understanding differences and ways of living 2. Consensual decision making 3. Interaction with space 4. Architectural program 5. Architectural schedule 6. Accessibility and connections	1. Capacity of multiple uses 2. Appropriation and adequacy of spaces 3. Percentage of areas of circulation 4. Quantity of spaces, accesses, and (types of) uses 5. Area/person ratio
Variable C: technique and technology	1. Ecotechnics 2. Acknowledgment and appreciation of the local 3. Comfort 4. Constructive system 5. Materiality 6. Project feasibility	1. Taking advantage of rainwater, solar energy, reusable waste and local materials 2. Taking advantage of local practices, vernacular techniques and knowledge 3. Thermic insulation, resistance and structural rigidity 4. Quantity of used material and costs
Variable D: sense and meaning	1. Encouraging autonomy 2. Co design participation 3. Education and local knowledge dialogue 4. Vision of future 5. Appropriation of space 6. Location of design	1. Self-governance methods 2. Quantity of consensual decisions 3. Implemented pedagogical practices 4. Interaction of user space 5. Collaborative design practices (quantity, depth, complexity, continuity, etc.)

2.7 Partial Conclusions

Both architects and designers can contribute to address the problems of everyday life in a strategic and flexible way, transforming communities into permanent laboratories for experimenting with forms of social innovation [45]. For this, new methods promote the exchange of knowledge and inter- or transdisciplinary practices in all phases of the design process in pursuit of a common goal. This implies complementing the conception of the architectural project only from the spatial, functional and technical aspects, to also think from the social and political responsibility of architecture based on the integral recognition of the environment, the ways of living of the inhabitants, the knowledge present in the territory and the dreams and purposes that the subjects have or may have in the complex scenario of overcoming an armed conflict.

Consequently, the architectural design for transition is eminently participatory and transdisciplinary in order to generate alternatives from the understanding of the ruptures in the problematic contexts. It is also presented as a tool with great potential to contribute to the profound ecological and cultural transformations and transitions emerging in the territories of difference, made up of places in which the logics of the communal are the very root of design [17]. Thus, the creative capacities to design and solve different situations typical of the everyday life of the inhabitants are understood and an architecture for peace is configured that values all the efforts of local populations, which are more transcendental in practice than public policies that seek to transform reality [73].

Architectural design in a scenario of transition from armed conflict to peace appears then as an instrument for the creation of spaces to live, work and meet, contributing to the non-repetition of the conflict from the promotion of autonomy, good living, and life in community. The theoretical model elaborated and presented in this chapter will allow directing the approach used to propose, develop and evaluate the architectural projects that aim to contribute to peace building, not only from what was agreed in the Final Agreement but also from those grassroots practices that will continue to emerge as the communities become stronger in their autonomy and in their vision of living in peace. In this sense, the model facilitates the review of the projects in the study area that will be presented in the following chapter and justifies the need to identify design tools that allow the inclusion of the variables and parameters presented here in future architectural projects within the framework of peace building in Colombia or in other similar contexts, mainly in the global south.

References

1. Acosta S, Diaz Diaz F (2018) Strategy for the sustainable use of ecotourism in the route of the colombian piedemonte. Europ J Sustain Dev 7(3):225–236
2. Agencia de Renovación del Territorio (2022) Agencia de Renovación del Territorio. Obtenido de Agencia de Renovación de Territorio
3. Agencia para la Reincorporación y Normalización (2021) Ruta de reincorporación social y económica. http://www.reincorporacion.gov.co/es.
4. Aschner JP (2019) Poetical, polite, political architecture in Latin America. In: IOP conference series: materials science and engineering, IOP Publishing, Praga, p 471. https://doi.org/10.1088/1757-899X/471/7/072026
5. Ayala ET, Rodriguez RD, Osorio EG (2016) Arquitectura y arte como medios para construir y fortalecer la Colombia del postacuerdo. Dearq (18):46–55. https://doi.org/10.18389/dearq18.2016.04
6. Barthes R (1986) Lo obvio y lo obtuso. Imágenes, gestos, voces. Ediciones Paidós, Barcelona
7. Berry T (1999) The great work: our way into the future. Bell Tower, New York
8. Boano C (2011) 'Violent spaces': production and reproduction of security and vulnerabilities. J Archit 16(1):37–55

9. Centro Nacional de Memoria Histórica (2016) Centro nacional de memoria histórica. Obtenido de Colombia, una esperanza para el mundo. http://www.centrodememoriahistorica.gov.co/micrositios/fin-del-conflicto/index.html
10. Ching F (1982) Architecture. Form, space & order. Gustavo Gili, Barcelona
11. Cigarini T (2017). Leer bajo la sombra. El colegio Schorge en Burkina Faso y el Plan Selva en el Perú. Arquitextos 24(32):39–56
12. Coalición Internacional para el Hábitat—oficina para América Latina HIC-AL (2017) Utopías en Construcción. Ediciones Navarra
13. De Carlo G (1980) An architecture of participation. Perspecta 17:74–79
14. di Campli A (2021) La Differenza Amazzonica. Forme ed ecologia della coesistenza. Letteraventidue edizioni, Siracusa
15. Diaz Diaz F (2022) Diseño arquitectónico para las transiciones. El posAcuerdo de paz en Colombia. Bitácora Urb Territ 32(1):121–134
16. Escobar A (2005) Más allá del tercer mundo. Globalización y diferencia. Instituto Colombiano de Antropología e Historia ICANH, Bogotá
17. Escobar A (2019) Autonomía y Diseño: La realización de lo comunal. Editorial Universidad del Cauca, Popayán
18. Espinosa N (2009) El campo jurídico del "otro derecho" en la Sierra de La Macarena. Elementos para un análisis crítico de la justicia alternativa en una zona de fuerte conflicto armado. Rev Colomb Sociol 32(2):133–155
19. European Committee of the Regions (2022) The EU's assembly of regional and local representatives. Obtenido de Recovery in rural areas can only happen by working together on their needs. https://cor.europa.eu/en/news/Pages/recovery-rural-areas.aspx
20. Fals Borda O (1968) The significance of guerrilla movements in Latin America. Cross Curr 18(4):451–458
21. Fals Borda O (2008) Detener la guerra. El Heraldo
22. Fitz A, Krasny E (2019) Critical care: architecture and urbanism for a broken planet. MIT Press Direct
23. Forjando Paz (2017) Ruta para la construcción de una paz estable y duradera. Torreblanca Agencia Gráfica, Bogotá
24. Fry T (2015) Design: on the question of "the imperative." Des Cult 7(3):417–422. https://doi.org/10.1080/17547075.2015.1105713
25. Fuerzas Armadas Revolucionarias de Colombia–Ejército del Pueblo FARC-EP (1998) FARC-EP Esbozo histórico
26. Fundación Guillermo Toriello (2004) World habitat. Obtenido de world habitat: https://www.world-habitat.org/wp-content/uploads/2016/03/FGT-Estudio-de-Impacto.pdf
27. García-Reyes MI, Anzellini S (2019) Saberes compartidos del hábitat: una arquitectura para el paisaje rural. Dearq (24):34–47. https://doi.org/10.18389/dearq24.2019.03
28. Garzón MA (2008) El lugar como política y las políticas del lugar. Signo Y Pensam 27(53):92–103
29. Giraldo Moreno SJJ (2015) Aportes sobre el origen del conflicto armado en Colombia, su persistencia y sus impactos. Espacio Crítico, Bogotá
30. Gobierno Nacional de Colombia & FARC-EP (2016) Acuerdo final para la terminación del conflicto y la construcción de una paz estable y duradera, de noviembre 24 de 2016. Mesa de Conversaciones
31. Goodman R (1971) After the planners. Simon and Schuster, New York
32. Gudynas E, Acosta A (2011) La renovación de la crítica al desarrollo y el buen vivir como alternativa. Utopía Y Prax Latinoam 16(53):71–83
33. Hábitat para la Humanidad (2020) Hábitat para la humanidad. Obtenido de La necesidad de vivienda en El Salvador. https://www.habitat.org/lac-es/where-we-build/el-salvador/
34. Hosie S (2016) La casa del pueblo: nuevo centro de El Salado. Dearq (19):104–111
35. Huizer G (1975) Organizaciones campesinas y su potencial revolucionario: estudios comparativos de algunos casos en América Latina y Asia. Estud Asia Y África 265–292

36. Irwin T (2015) Transition design: a proposal for a new area of design practice, study, and research. Des Cult 7(2):229–246
37. Isaak C (2016) Sobre la memoria y la arquitectura: construir la ausencia. Dearq (18):80–87. https://doi.org/10.18389/dearq18.2016.07
38. Kay C (2007) Enfoques sobre el desarrollo rural en América Latina y Europa desde Mediados del Siglo Veinte. In: Perez E (ed) La enseñanza del desarrollo rural: enfoques y perspectivas. Pontificia Universidad Javeriana, Bogotá, pp 49–111
39. Kéré DF (2017) Entrevista a Francis Kéré, Constructor comunitario. In: Martinez de la Vega A (ed) Entrevistador
40. Latour B (2018) Down to earth social movements: an interview with Bruno Latour. Denise Milstein IM-G-G (ed) Entrevistador
41. Lizondo Sevilla L (2011) La función en la arquitectura. In: Andar por casa. En torno al análisis del proyecto. General Ediciones de arquitectura, pp 51–57
42. Lopez RO, Quinteros C, Ramos CG (2015) Inclusive political settlements. Obtenido de Reforma del Estado después de los Acuerdos de Paz Negociando e implementando un acuerdo politico inclusivo en El Salvador. https://ips-project.org
43. Loren-Mendez M (2018) Bernard Rudofsky: architecture without Architects, a short introduction to non–pedigreed architecture. Proy Prog Arquit 120–121
44. Magnaghi A (2010) Il progetto locale. Bollati Boringhieri Editori, Torino
45. Manzini E (2015) Design, when everybody designs: an introduction to design for social innovation. MIT Press, Cambridge
46. Mendoza C (2001) Universidad Rafael Landivar. Obtenido de Universidad Rafael Landivar. http://courseware.url.edu.gt
47. Mezones F (2005) Obtenido de Escuela Superior de Administración Pública. https://repositoriocdim.esap.edu.co/bitstream/handle/123456789/8844/6587-1.pdf?sequence=1
48. Ministerio de Agricultura y desarrollo rural (2017) Decreto 893. Bogotá
49. Molano A (1987) Selva adentro. Una historia oral de la colonización del Guaviare. El Áncora editores, Bogota
50. Molano A (2000) The evolution of the FARC: a guerrilla group's long history. NACLA Rep Am 34(2):279–283
51. Montobbio M (1997) JSTOR. Obtenido de JSTOR. https://www.jstor.org/stable/20644189
52. Morton T (2007) Ecology without nature: rethinking environmental aesthetics. Harvard University Press
53. Ortiz C (2003) Cultivos ilícitos y nueva ruralidad en Colombia. Cuad Desarro Rural (50):143–166
54. Paba G (1994) Città di pensiero e di azione: principi di ricerca del laboratorio di progettazione ecologica degli insediamenti. Bollettino del Dipartimento di Urbanistica e Pianificazione del Territorio, Università degli studi di Firenze, pp 12–15
55. Peñaranda Currie I (2021) L'eredità spaziale del fallimento. Un'analisi dei programmi di colonizzazione statale nel dipartimento di Caquetá, Colombia (1959–1981). In: Di Campli A (ed) La differenza amazzonica. Forme ed ecologie della coesistenza. LetteraVentidue Edizioni, Siracusa, pp 54–69
56. Pérez CE (2001) Hacia una nueva visión de lo rural. In: Giarracca N (ed) Una nueva ruralidad en América Latina? CLACSO, Bueno Aires, pp 17–29
57. Pérez H, León F, Marmolejo L, Rodríguez J, Claudia G, Ruecker S, Fajardo F (2018) Diseño para la reconciliación: El co-diseño de un futuro pacífico en zonas de posconflicto en Colombia. Diseña 13:140–173
58. Petrescu D, Trogal K (2017) The social (re)production of architecture: politics, values and actions in contemporary practice. Routledge, London
59. Programa de las Naciones Unidas para el Desarrollo (2011) Informe Nacional de Desarrollo Humano 2011. Colombia rural, Razones para la esperanza. PNUD, Bogotá
60. Raskin P (2002) great transition: the promise and lure of the times ahead. Stockholm Environment Institute

61. Rincón JJ (2006) Movilización social y desarrollo rural en Colombia: del sistema agrícola a la nueva ruralidad (1990–2002). Rev Colomb Sociol (27):63–98
62. Rogers P, Ramsbotham O (1999) Then and now: peace research past and future, political studies. Polit Stud 740–754
63. Ruiz Serna D (2003) Campesinos entre la selva, invasores de reservas. Tabula Rasa (1):183–210
64. Sala Valdés C, Barros N, Salamanca MD, Álvarez M, Iokiñe R, Uribe JJ, Santiago P (2022) Paces locales y estrategias transformadoras: más allá de la paz dominante en el sur del Tolima. In: Uribe JJ, Rodriguez I, Baquero J (eds) Paces desde abajo. Desafíos y oportunidades de otra paz. Universidad del Rosario, Bogotá, pp 91–115
65. Santos BD (2006) Capítulo I. La sociología de las ausencias y la sociología de las emergencias: para. In: Santos BD (ed) Renovar la teoría crítica y reinventar la emancipación social (encuentros en Buenos). Buenos Aires, pp 13–41
66. Santos BD (2011) Epistemologías del sur. Utopía Y Prax Latinoam 16(54):17–39
67. Schneekloth L, Shibley R (2000) Implacing architecture into the practice of placemaking. J Archit Educ 53(3):130–140
68. Sepúlveda JG (2012) Barrios populares: hacia la búsqueda de la producción social del hábitat en Bogotá. Bitácora Urbano Territ 1(20):145–156
69. Speth JG (1992) The transition to a sustainable society. Proc Natl Acad Sci USA 89(3):870–872
70. Till J (2005) The negotiation of hope. In: Blundell Jones DP (ed) Architecture and Participation. London, pp 25–44
71. Unidad para la Atención y Reparación Integral a las Victimas (2022) Registro Único de Victimas. Bogotá, Colombia
72. Unigarro Caguasango DE (2020) Los campesinos de la Amazonia noroccidental colombiana: entre la coca, el conflicto y la construcción de paz. Antípoda Rev Antropol Y Arqueol (40):175–200
73. Unigarro D, Sanabria L, Bonilla L (2021) Indagación de la memoria, reconciliación y arquitectura para la paz en Monterrey (Casanare). In: Rojas J, Vargas A (eds) Territorio, equidad y desarrollo. Ediciones Unisalle
74. Unión Europea (2022) Feeding Europe. 60 years of common agricultural policy. Obtenido de La política agrícola común en pocas palabras, Luxemburgo, Unión Europea
75. URNG (2004) Obtenido de memoria virtual guatemala. https://memoriavirtualguatemala.org/
76. Velásquez CA (2018) Unstructured interview. (F. Diaz Diaz, Interviewer)
77. Velez MA (2001) FARC–ELN: Evolución y expansión territorial. Desarro Y Soc (47):151–225
78. Vergara W (2011) Desarrollo del subdesarrollo o nueva ruralidad para Colombia. Cartografías del desarrollo rural. Rev Univ Salle (55):33–66
79. Vilas CM (1996) PAPERS. Obtenido de Un balance de la ejecución de los acuerdos de paz en El Salvador. https://papers.uab.cat/article/view/v49-vilas/pdf-es
80. White D (2020) Just transitions/design for transitions: preliminary notes on a design politics for a green new deal. Capital Nat Soc 31(2):20–39
81. Willis A-M (2006) Ontological designing-laying the ground. Des Philos Pap 4(2):69–92. https://doi.org/10.2752/144871306X13966268131514
82. Yagenova SV (2017) JSTOR. Obtenido de JSTOR. https://www.jstor.org/stable/j.ctv253f5v7.25
83. Yuste M (1999) La ex Guerrilla de Guatemala pide perdón por sus actos violentos. El Pais Dig
84. Zuleta Ruiz B, Torres Tovar C (2022) Ciclos y tendencias de los hábitats de la guerra en la colonización territorial en Colombia. Bitácora Urbano Territ 1(32):7–15

Chapter 3
The Architectural Project in the Transition to Peace in the Case Study

Abstract This chapter aims to approach and review the architectural projects carried out in Colombia, particularly in the municipality of San José del Guaviare, in the framework of the implementation of the Peace Agreement signed in 2016. For that, a contextualization of the programs and public policies that have allowed the management, design, and construction of these projects at the national level will be carried out first. In a second moment, the experiences, methodologies and instruments of the workshops and visits that were developed in the fieldwork phase of this research will be described. Finally, some architectural projects in the study area will be carried out using the theoretical model presented in the previous chapter as an evaluation tool. In this way, it is intended to show findings and useful contributions for the consolidation of the design tools for the transition to peace that will be presented in the last chapter.

3.1 Projects Derived from the Peace Agreement in the National Context

The signing of the Peace Agreement in November 2016 between the national government and an illegal armed group that was active for more than 60 years, the Revolutionary Armed Forces of Colombia—FARC-, allowed the generation of instruments and public policies to favor its implementation. The two main project scenarios in this regard have been mentioned in previous sections and will be described and analyzed in greater detail in this chapter: the Territorial Focused Development Plans -PDET- and the Territorial Spaces for Training and Reincorporation—ETCR-.

These instruments are part of a process of territorial transition in two senses. First, a transition towards peace, given that the internal armed conflict experienced by the country since the middle of the last century marked the future of the Colombian rural territory not only due to countless violent actions, but also as a place of deprivation, marginalization, and state neglect. In view of this, the signing of the Final Agreement lays the foundations for a peace-building process aimed at reconciliation and overcoming violence. Second, a transition towards autonomous and sustainable ways

F. R. Diaz Diaz, *Architecture and Peacebuilding*, PoliTO Springer Series,
https://doi.org/10.1007/978-3-032-13585-8_3

of life in rural territories with the support and the participation of all institutional, social, and corporate actors for the generation of possible and viable alternatives that include strategic actions and projects for both management and intervention to improve the conditions of existence and guarantee the "good living" of the communities. In practical terms, this includes the development of infrastructure projects, community facilities and rural housing, in which architectural design must transcend its technical character to become an element of innovation and social transformation, an important device for territorial transition [7].

3.1.1 PDET

The Territorial Focused Development Programs—PDET- were created by Decree 893 of 2017 and have enabled the execution of more than 2,500 works as of February 2023 aimed at improving community, educational, road and basic sanitation infrastructure [1]. Its scope focuses on 170 municipalities prioritized for having a high incidence of multidimensional poverty, a high incidence of armed conflict and a high concentration of coca crops [12]. The PDET have been defined by the Agency for Territorial Renewal (2021) as special instruments for "15-year planning and management, which aim to stabilize and transform the territories most affected by violence, poverty, illicit economies and institutional weakness, and thus achieve the rural development required by these 170 municipalities". These in turn have become a mechanism for the implementation of the Comprehensive Rural Reform agreed in the Final Agreement, as it seeks to contribute to the "structural transformation of the countryside, create welfare conditions for the rural population, protect the multiethnic and multicultural wealth, promote the peasant economy, integrate the regions affected by violence, strengthen community organizations and turn the countryside into a scenario of reconciliation" [1].

The importance of these territories lies not only in the impacts and vulnerability described above, but also in the fact that they constitute 36% of the national territory where more than six million Colombians reside and 24% of the country's rural population, of which 57% is rural population with multidimensional poverty. Likewise, more than half of the victims of the armed conflict lives in these territories [1].

The project structuring process was carried out in phases. The first phase, developed in the hamlets, was led by the Agency for Territorial Renewal -ART- and community delegates who integrated the working groups that prioritized the initiatives from the voices of the local inhabitants, establishing the Community Pacts for Regional Transformation—PCTR-. These were taken to a second phase at the municipal level, where the delegates participated along with representatives of social organizations, public and private entities to establish the Municipal Pacts for Regional Transformation—PMTR- which were taken to the next stage. In this last phase, the 170 municipalities were grouped into 16 sub-regions (Fig. 3.1), establishing for each sub-region the Action Plans for Regional Transformation—PATR—which

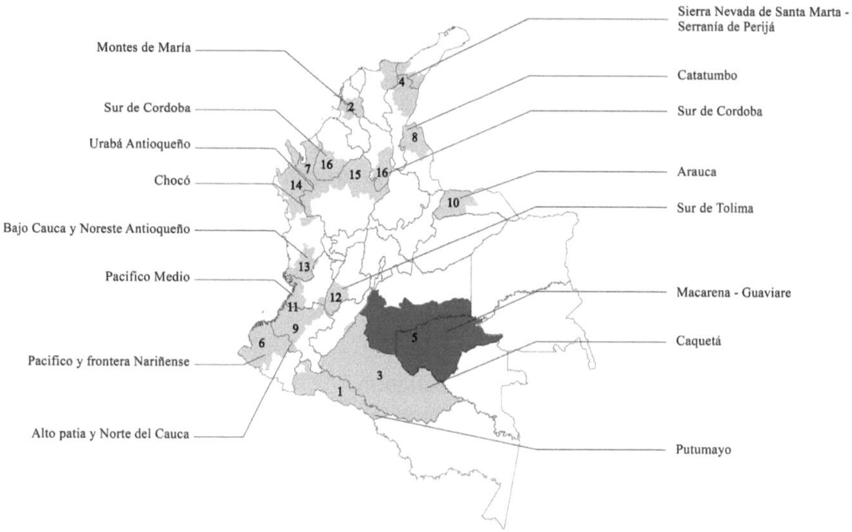

Fig. 3.1 PDET subregions location. By the author, 2022

included about 33,000 initiatives and were the basis for the PDETs, with a 15-year implementation goal and updated every five years.

The peace agreement in Colombia integrates part of the arguments that led to the structuring of the PDETs in the proposal for the Comprehensive Rural Reform, the first item on the thematic agenda of the negotiating table. It proposes to "guarantee the development of the peasant and family economy as the basis for the country's food sovereignty" [8], as an alternative to solve rural poverty and the predominance of agribusiness and monocultures. The Comprehensive Rural Reform also recognizes the inequitable distribution of land as a problem to be solved in the implementation of the final agreement, through the "democratization of access to land for the benefit of rural communities most affected by poverty, state neglect and conflict" [8]. The Government and the FARC-EP set the common goal of "contributing to the structural transformation of the countryside, closing the gaps between the countryside and the city, creating conditions of well-being and good living for the rural population" [10].

This highly participatory scheme represented an unprecedented design exercise in the country in pursuit of comprehensive rural reform. The effort to prioritize 33,000 initiatives proposed from the grassroots level has meant that many of the projects built or executed have been easily accepted by the communities that proposed them at the time. However, during the implementation stage, spaces for participation have disappeared, which poses new challenges for ownership in the medium and long term. Some of the reasons for the decrease in community participation have to do with the technical approach to implementation, in which the territorial entities that execute them gain prominence, leaving the grassroots groups behind. Likewise, the characteristics of the contracts with which most of the works are assigned require reduced deadlines that do not facilitate socialization, discussion, and adjustments during the process. Some communities also reduced opportunities for socializing and get together to the pandemic experienced during 2020.

PDETs as a planning and management instrument to implement the different plans and projects derived from the Final Agreement prioritize the component of social and community projects [8]. To categorize the initiatives and works of these programs, eight pillars were established that have been the structure of the interventions in the subregions:

1. Social management of rural property and land use
2. Infrastructure and land development
3. Rural health
4. Education and rural childhood
5. Rural housing, drinking water and sanitation.
6. Economic recovery and agricultural production
7. Right to food
8. Reconciliation, coexistence, and peace

Table 3.1 shows the number of PDET works that have been executed for each pillar in each subregion as of 1st March 2023. The importance of these works for reincorporation and peace building in the prioritized territories lies mainly in the collective and participatory promotion with which they are managed. Likewise, the articulation with the actions carried out in the framework of the reincorporation policies favors the sustainability of progress and peace building. This is particularly because in 12 of the 16 sub-regions, at least one ETCR is present [1]. On the other side, the implementation of productive projects in this framework promotes the consolidation of legal and local economies that contribute to the eradication of the causes that originated and possessed the conflict. This is the measure of the importance of PDET in the transitional process of peace building in Colombia, and the architectural projects that are part of this program cannot be alien to the objectives of the local inhabitants. In order to respond to this priority approach, funding is obtained from the following sources summarized in Table 3.2.

- OCAD-PAZ: Collegiate Administration and Decision for Peace, which obtains resources from the General Royalty System. It is the largest pool of resources for peace to which any type of project prioritized in the PATR can have access.
- PDET-ART: Obtains resources from the General Budget of the Nation, allowing the allocation directly from the state entities to facilitate the execution of priority works in the different pillars through the Territorial Renewal Agency and other territorial public entities.
- Works for taxes: Private companies allocate resources for the execution of PDET projects in exchange for tax benefits.
- Private projects: Resources obtained from private companies and individuals.
- International cooperation: Resources mainly from the United Nations Multi-Donor Fund, European Union Trust Fund, World Bank Fund (loan), USAID, German Development Bank (loan) and cooperation agencies of guarantor countries.

Table 3.1 PDET works by pillar in every subregion. By the author based on official data, 2023

Subregion		Quantity of completed PDET works								
Pillar		1	2	3	4	5	6	7	8	Total
1	Alto patía y norte del Cauca	1	132	2	111	16	34	0	69	365
2	Arauca	4	52	6	18	3	5	0	12	100
3	Bajo Cauca y nordeste antioqueño	7	55	6	105	26	26	0	115	340
4	Catatumbo	7	94	5	78	9	10	0	27	230
5	Chocó	2	31	1	24	3	11	1	36	109
6	Cuenca del Caguán y piedemonte caqueteño	2	142	0	24	7	10	1	23	209
7	Macarena–Guaviare	1	130	1	81	27	9	1	57	307
8	Montes de María	1	36	1	30	3	10	0	22	103
9	Pacífico medio	2	2	0	0	0	5	0	12	21
10	Pacífico y frontera nariñense	3	54	10	55	4	26	0	36	188
11	Putumayo	2	104	4	61	47	41	0	83	342
12	Sierra Nevada–Perijá	2	59	2	40	7	28	1	16	155
13	Sur de Bolívar	2	28	0	13	1	9	0	10	63
14	Sur de Córdoba	2	35	0	33	3	15	0	10	98
15	Sur del Tolima	1	35	1	33	0	9	0	18	97
16	Urabá antioqueño	1	15	1	44	4	14	0	20	99
	Total	40	1004	40	750	160	262	4	566	2826

Table 3.2 Source of financing projects PDET. By the author based on ART data, 2023

Source of financing	Number of projects	Value in COP	Value in USD
OCAD-PAZ	159	$ 915.458.380.000	$ 193.542.998
PDET-ART	2.033	$ 465.347.390.000	$ 98.382.112
Works for taxes	60	$ 414.560.370.000	$ 87.644.899
Private projects	104	$ 7.894.600.000	$ 1.669.049
International cooperation	285	$ 121.143.710.000	$ 25.611.778

The PDETs have been the main scenario for the management, design and construction of architectural projects that contribute to peace through collective reparation and the generation of community spaces that promote social encounter and well-being in the rural areas of the municipalities most affected by the conflict. This instrument has also made it possible to show that the transitional process towards peace is not only a legal and juridical problem, but that peace is built from below, from the point of view of the local inhabitants who promote good living through daily actions. "The operational links between space, protection and vulnerable spatialities that emerge after a conflict or disaster, as well as the discourses around the notion of protection, tend to focus on legal issues of rights and responsibilities rather than on issues of physical and social protection that may depend on spatial relations" [3].

3.1.2 ETCR

Point three of the Final Agreement is focused on the end of the conflict. For that, a first phase aimed at the surrender of weapons by the FARC-EP by means of a technical procedure through which the ONU received the totality of the armament surrendered and destined it for the construction of three monuments [8]. To guarantee this process and the reincorporation of the guerrillas to civilian life, the guerrillas and the government established seven Transitory Normalization Points (Puntos Transitorios de Normalización–PTN-) and 20 Transitory Zones for Normalization (Zonas Veredales Transitorias de Normalización–ZVTN-), which had the function of being the first place of concentration and demobilization of the combatants. The location of these areas can be seen in Fig. 3.2.

These PTNs and ZVTNs had a transitory character which ended on August 15, 2017, date on which were constituted in some of these same areas, the Territorial Spaces for Training and Reincorporation -ETCR-, temporary rural settlements designed to host training and early reincorporation processes to facilitate the adaptation of FARC-EP members to civilian life [2]. The ETCRs, in addition to including temporary housing solutions, were considered places where to carry out "training and early reincorporation activities with which it is intended to facilitate the initial phases of adaptation of FARC-EP members to civilian life" [2].

This legal figure, also transitory, ended on August 15, 2019. Since then, these spaces entered a transition process under the figure of Former ETCR (Antiguos ETCR)—AETCR-, with the goal of becoming Populated Centers—rural territorial figure, officially recognized as having at least 20 grouped dwellings-, or being relocated or intervened. 11 of the 24 sites are in the process of relocation due to natural risks, limited access roads and public services, among other factors. The remaining 13 in the process of transitioning their territorial status to that of a population center, which will allow them to legalize their cadastral situation and officially connect to the municipality where they are located [2]. The location of the 24 ETCR and the identification of those in the process of relocation and consolidation can be seen in Fig. 3.3.

These two instruments, PDET and ETCR, generated new rural architectural typologies based fundamentally on prototypical and industrialized solutions. In a way, the community centers that have been built within the framework of the PDET, an example of which can be seen in Fig. 3.4, and on the other side rural housing models for the ETCRs, such as the one shown in Fig. 3.5. However, these spaces

Fig. 3.2 PTN and ZVTN location. By the author, 2022

have not had the expected results due to different technical, constructive, management and acceptance difficulties on the part of the user communities, which have not allowed their correct appropriation.

This proposes a challenge and a unique opportunity for architects to act for the transition to the objectives related to peace building. In this context, design variables and tools different from those used in other design processes must be considered given the special sensitivity and historical importance of these practices. "Designers who believe they are engaged in the architecture of protection, delivering shelters or segmenting spaces, may be involved in a different form of violence and coercion, imposing a new disciplinary gaze" [3].

Fig. 3.3 ETCR location and intervention process. By the author, 2022

Piquard and Swenarton [14] address the discussion on the role of architecture in post- war processes by arguing that "working on architecture and space alone cannot reduce or transform conflict, but understanding the spatial dimension of conflict can be crucial to a comprehensive response to it" [14]. Along these lines, the practice of architectural design within the conditions to guarantee stable and lasting peace implies necessarily aiming at the recovery of its social function from the understanding of the spatial dynamics associated with the post-agreement.

Fig. 3.4 Community center in San José del Guaviare. By the author, 2021

Fig. 3.5 Housing in the ETCR Las Colinas, San José del Guaviare. By the author, 2021

3.2 Methodology, Instruments, and Findings from Fieldwork

As explained in previous sections, San José del Guaviare is the object of study of this research due to its geopolitical importance during the conflict, which has been reflected in the considerable presence, and above the national average, of architectural projects in the framework of the implementation of the Peace Agreement: the only municipality with two ETCR and 42 PDET works, including community center, sanitary batteries, road improvement and improvement of educational facilities. The territorial extension of San José del Guaviare is 16,178 km2 which are divided into 84 rural territorial areas called *veredas*. The *vereda* is a subdivision structure of the municipalities used in Colombia to designate those settlements with a wide rural extension that may or may not have villages located on the roads that connect two municipalities. For the purposes of the English language version of this monograph, the term **hamlet** will be used to refer to ***vereda***.

In order to conduct a reconnaissance of these projects and obtain primary source information for the development of the second specific objective, five field visits were conducted between 2018 and 2021. The methodology, instruments and findings of these visits will be presented below.

3.2.1 International Workshop "Pensando el Territorio Veredal"

The methodology used for this research is based on participatory exercises with rural communities in the municipality of San José del Guaviare carried out privileging the empirical method of participatory action research. These practices have their initial scenario in the International Workshop "Pensando el Territorio Veredal"—Thinking the Hamlet Territory—in its three versions carried out between October 2018 and October 2019. This Workshop was organized by the Universidad de La Salle, Bogota of Colombia and the Università degli Studi G. d'Annunzio Chieti-Pescara of Italy. Several local ONGs, community associations and the inhabitants of the different hamlets also contributed to this experience.

The workshop was held in the hamlets of Charras and Charrasquera. They are located 90 km northeast of the town of San José del Guaviare in a journey of approximately three hours in a 4 × 4 vehicle on an unpaved road. During the tour, the landscape is transitory, and the jungle becoming increasingly dense and the houses more isolated. This road was marked by the first migrations of settlers who arrived in San José del Guaviare in the 1950s and 1960s fleeing the bipartisan war [13], but the settlements were founded during the 1980s as a consequence of the colonization wave resulting from the coca bonanza. During the 1990s and 2000s, with the presence of different armed groups vying for territorial control, they became the site of violent victimizing events that caused massacres and massive displacements. With the signing of the peace agreement, the inhabitants of these hamlets regained the right and confidence in their territory and were recognized as victims of the armed conflict subject to collective reparations.

The first version of the workshop, held in October 2018, was held in the hamlet of Charrasquera. At the time of this first workshop, there were only two houses in the village and there was no power, water or sewage service in the entire hamlet. In the surrounding area there were 45 families living in an equal number of scattered farms. The main objective of this workshop was to generate a multidimensional recognition of the territory useful to establish a diagnosis and accompany the peasant community in the process of structuring the Sustainable Development Plan of the hamlet. For this purpose, several activities were carried out based on the participatory design approach, identifying the needs and strengths of the local population in order to propose some general lines of action to structure the development plan prioritizing four components that the community recognized as heritage: (1) the natural heritage, (2) the human heritage, (3) the social heritage and (4) the physical heritage.

The activities were carried out in working groups where local inhabitants and students from the two universities discussed and represented the different territorial dimensions, establishing a participatory diagnosis. The instruments used were paper, markers and satellite maps, which made it possible to generate social cartography products that located the farms in the hamlet in relation to elements of the physical and spatial structure of the territory, such as pipes and roads. Collective mapping is a qualitative and participatory information gathering tool that will be presented in the fourth chapter of this research as one of the useful design tools in these contexts.

This meeting made it possible to demonstrate the knowledge and attachment of the inhabitants to their territory and to generate bonds of trust and mutual respect among the participants. Likewise, the generalized concern of finding an economic activity other than coca cultivation, which at the time allowed them to supply all their needs and even live with a certain ostentation.

The first version was also the opportunity to conduct a preliminary reconnaissance of the hamlet of Charras where the ETCR Marco Aurelio Buendía is located. This village is located on the banks of the Guaviare River, 20 km from Charrasquera. Its inhabitants also suffered displacement due to the conflict in 2002, because of the incursion of paramilitary groups that arrived in Guaviare in 1998, unleashing a war for the control of the territory. In this space a meeting took place where we listened to the ex-combatant leaders to understand how they organized themselves in the initial phase of reincorporation in this territory.

Some of the most outstanding findings of this first version point to three aspects: (1) the willingness of the inhabitants to adhere to illicit crop substitution programs included in the Peace Agreement and intended for local people to develop productive projects as an alternative to coca crops; (2) the optimism generated by the presence of the different public and private entities that allowed them to alleviate the feeling of loneliness and marginality that had characterized them during decades of conflict; and (3) the need to self-manage the present and the future once there are no longer armed groups telling them what to do.

Based on the trust generated with the community and the intention to continue the collaborative work between academia and inhabitants, it was proposed to hold a second version of the workshop in March 2019. On that occasion, the objective was to implement concrete actions aimed at the integral improvement of housing and the idea of ecotourism as a sustainable economic activity in the hamlets of Charrasquera and Charras. Returning to the territory in March 2019, five months after the first visit, the changes in the conditions of the place were evident: a community center had been built as part of the PDET works, there were three new houses in the village and individual solar energy systems had been installed in each house. For the first time in 17 years, the hamlet had electricity again.

The Charrasquera community center is in the main village of the hamlet. This place was the scene of several violent events that caused its destruction in 2002. Since then, the village has been slowly rebuilt, mainly after the signing of the Peace Agreement in 2016. Due to the disproportionate impact of the conflict, this place was included as a subject of collective reparation by the Victims Unit [16]. Consequently, and in line with the PDETs, the construction of a community center in this location was prioritized.

Its location is determined by the identification of a communally owned lot, located at the intersection of the two main roads. In the past, some community facilities such as the slaughterhouse were located there. Near the center, there are the ruins of a house destroyed in the 2002 victimization and some houses that have been rebuilt and inhabited again.

The construction of the center was executed by the United Nations Development Program (UNDP). Although the design and construction processes should have included the participation of the community, this participation did not take place, according to community members in different interviews conducted in March 2019. Thus, the construction techniques and architectural typology do not correspond to the ways in which the territory has been traditionally built and inhabited and to the way in which the hamlet has been rebuilt.

A first co-creation exercise allowed for the intervention of the recently built community center. The decision with the community was to build an outdoor kitchen since the kitchen included in the building did not correspond to their traditional ways of relating to this space. In addition, an opportunity was identified to exchange knowledge about materials and construction techniques and to apply them in the execution of an ecological stove based on raw earth blocks that would make visible a prototype that could be replicated in the houses. This stove, proposed by researchers and students at La Salle University, optimizes the consumption of firewood and reduces and channels the smoke to prevent it from being inhaled directly by the stove. This proposal arises as a response to a manifest problem of rural dwellers in Colombia who, due to their practices, have been exposed to respiratory diseases resulting from cooking with firewood.

For this purpose, an activity was carried out that allowed the local inhabitants to get closer to the raw earth as construction material. This activity took place in the ETCR of Charras, which was an opportunity for meeting and reconciliation between ex-combatants and peasants, once victimizers and victims. The collaborative work

Fig. 3.6 Intervention in the community center. By the author, 2019

around learning allowed to blur those social boundaries of the conflict and promote a space in which experts, peasant communities, students and ex-guerrillas contributed and learned. In the following days, the construction of the stove and the kitchen enclosure began with the active participation of the community (Fig. 3.6).

A second line of work in this version focused on ecotourism as a productive project proposed by ex-combatants in the ETCR. The proposal consisted of generating tours inside the jungle and the construction of replicas of the spaces they inhabited during the war. The goal of the working group was to act as test tourists, mapping the route with GPS instruments and making audiovisual recordings. At the end of the tour, there was feedback with observations and suggestions from the participants about the activity to identify areas for improvement.

Some of the most representative findings of this second version were focused on three aspects: (1) The community has the capacity to manage its own processes to appropriate spaces such as the community center; (2) local people have different knowledge that is enhanced when exchanges are generated with people outside the community; and (3) it is in the scenarios of collective construction where the opportunities to build peace from below, from everyday actions.

The third version of the workshop was held in October 2019 in response to the agreement with the community to continue accompanying their processes. The first finding was that the community continued the construction of the kitchen and made the enclosure with raw earth in one of the techniques shown in the second version named wattle and daub. In addition, for the third version, objectives focused on working with territorial memory and participatory ecological environmental monitoring were proposed.

An increase in the population of the village was evidenced. Between October 2018 and October 2019, new houses were built and several of the inhabitants who had been displaced during the conflict returned to recover their land. Likewise, we found that many inhabitants had privileged their personal or family projects over community initiatives. This was due to some failures experienced in crop substitution policies and projects promoted by the state, which revived the community's distrust of institutions.

The first activity of this version was to continue with the enclosure of the kitchen. The walls that had already been built by the community were plastered and painted. Likewise, the ecological stove was painted and plastered. In this way, the initial image and perception of the community center was modified, favoring the use and appropriation of the space by the inhabitants. A second activity focused on satellite mapping with GPS recording applications to locate some farms and their relationship with the environment. Finally, a third activity focused on re-signifying the ruins of an armed attack that took place in this hamlet in 2002. The objective was to build a community garden and paint a mural, to change the meaning that the inhabitants gave to this place of memory of the war.

The findings of this third version were relevant for the conclusion of the series of versions of the workshop: (1) The trust of rural communities towards institutions is fragile due to the scarce historical presence that has configured the marginality of these territories; (2) the sense that inhabitants give to a place is dynamic, so it must be consistent with the real needs of the community; and (3) the memory of the conflict must remain alive to promote non-repetition. However, from memory, spaces can be re-signified to give them a new sense aimed at the vision of the future.

The workshop *"Pensando el Territorio Veredal"* and its three versions were a participatory research scenario that allowed reflection on the role of the university in the peace-building process. The activities carried out and the methodology used allowed the collection of information not only for this research but also for the work of other researchers and students who in recent years have produced undergraduate theses, articles, book chapters, among others. Several of these publications are being cited in this work and constitute a set of contributions towards the visibility and discussion of peace building processes associated with the understanding of space.

3.2.2 *Fieldwork 2021–2022*

Due to the pandemic of 2020, the visits scheduled for that year were postponed and could only be resumed in August 2021. On this occasion, the objective regarding this work was to visit the two AETCR and the 17 hamlets where there is community center built in the framework of the PDET in San José del Guaviare. Other PDET works that have been built in these hamlets were also identified. Figure 3.7 and Table 3.3 show the location of the community centers visited.

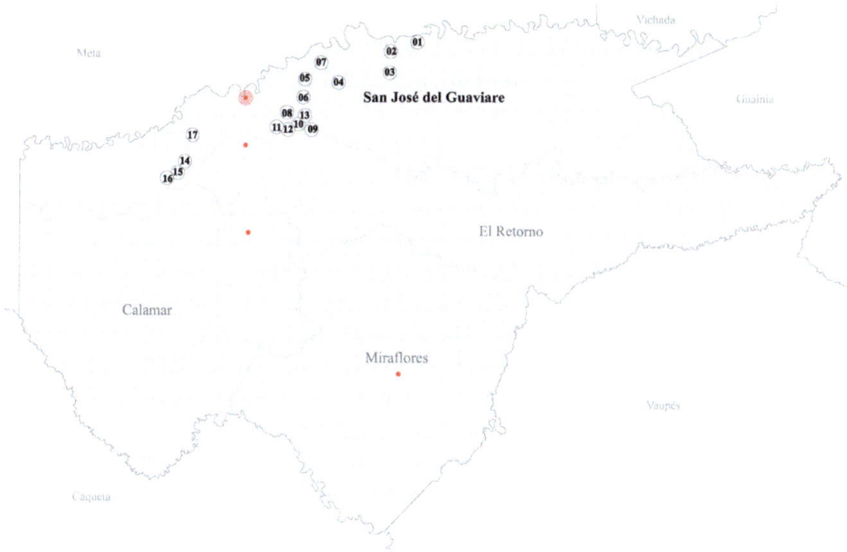

Fig. 3.7 Location of the community centers in San José del Guaviare. By the author, 2022

Table 3.3 List of the community centers in San José del Guaviare. By the author, 2023

Hamlet/PDET Community Center	
1	Unión de Buenos Aires
2	Charras
3	Charrasquera
4	San Luis
5	Las Dunas
6	El Morro
7	Sabanas de la Fuga
8	San Cristóbal
9	Caño Mosco
10	Santa Cecilia
11	Baja Unión
12	La Dosmil
13	Puerto Ospina
14	El Rosal
15	Caño Lajas
16	Caño Pescado
17	Las Orquídeas

Different instruments were used to collect information for this field trip. Firstly, a drone was used to record aerial photos and videos of the community centers to identify the relationship between the buildings and their surroundings. Secondly, the MyTracks application was used to record by GPS the movements between the places visited. Finally, an audio recorder was used to collect unstructured interviews with local residents to obtain their perceptions of these projects.[1]

This tour through the different hamlets revealed relevant findings for this research related to the conditions in which these spaces were managed, designed, and built. In the first place, with respect to the 17 centers, a first finding was the almost exact coincidence in the designs of all the buildings regardless of the differences in the environments where they are located. The use of materials such as concrete and brick was also recurrent. In no case were local materials considered despite the presence, in some cases, of nearby community constructions made of wood, and the construction culture of this region of the country based on jungle materials.

The presence of these 17 centers has also generated distrust regarding the management and contracting processes of the works. Although their management was required by the community through the participation instruments of the Community Pacts for Regional Transformation, there was not enough community participation in the execution of resources and the selection of contractors. In some cases, local labor was not considered even though it was one of the conditions agreed upon in the PCTRs. The community centers have also been a symbol of collective reparation and state presence. However, this presence has been gradually reduced and the communities have had to continue their processes autonomously. This is where architecture can detonate local initiatives that lead to the encounter and community building that will allow progress towards peace.

This field visit also allowed the recognition of the two AETCR. On the one hand, the AETCR Marco Aurelio Buendía, located in the hamlet of Charras, is in the process of relocation since August 2019. This situation has made it evident the lack of sense of belonging to the place they occupied at the time of the visit. The number of permanent inhabitants was lower than that found in 2018 and those who remain there, do so while waiting for a definitive solution to the relocation with a comprehensive improvement of their quality of life.

In contrast, the AETCR Jaime Pardo Leal of hamlet Las Colinas is in the process of consolidation as a populated center. The slow but successful result of the acquisition of the land, as well as the arrival of new people to the settlement and the growth of productive projects have made this community consolidate and advance in a self-managed way in its processes. During this visit, unstructured interviews were conducted with different leaders of the settlement to identify relevant aspects of the reincorporation process they are carrying out.

AETCR Charras—"Noble and Peaceful Community Marco Aurelio Buendía"

[1] The information collected through these instruments is available at https://sites.google.com/view/toolsfortransition/home.

On the border between the department of Meta and Guaviare and approximately four hours from San José del Guaviare, is located the AETCR of Charras, which is established as one of the most important reincorporation spaces of the post-agreement due to its location in what was one of the most conflictive zones during the war. It has a health center, a library, several leisure areas and 55 housing modules with five rooms each, with construction materials supplied by the state. Its name honors the FARC leader alias Marco Aurelio Buendía, who operated repeatedly in the departments of Meta, Boyacá and Cundinamarca, the so-called "Eastern Bloc", and who became a trusted member of the FARC secretariat.

The AETCR was officially established in January 2017 with approximately 250 ex-combatants coming from various guerrilla fronts in the country and their families that by 2022, were reduced to about 127. The land planned for the settlement is not arable since it belongs to the Amazon Forest Reserve Zone, limiting its productivity and giving rise to a resettlement planned by the national government. A 197-hectare property is acquired in the nearby village of El Boquerón in 2022, to give continuity to the reincorporation process and agricultural productivity [2].

The land of the current AETCR consists of 10 hectares initially under lease until December 2020. It was finally acquired by the community in April 2021. The allocation of the land is foreseen on an individual basis per household and by consensus, it is determined that the remaining land will be used for the development of productive projects, specifically, an agro-ecological cooperative, a nature tourism project, a productive-environmental training school and several collective agricultural initiatives that supply the community itself and a supermarket in San José del Guaviare.

Although the community has services such as gas supply, a drinking water treatment plant and energy through electric generators, there is a lack of sewerage, water supply networks and constant internet connectivity, since there are two antennas financed by the community and a learning space with 20 computers. This is part of several self-managed spaces modified by the community, to which are added several modules for social gatherings, a shuffleboard court, a community vegetable garden, two nurseries for rent and a community store. Most of them are built with a wooden structure, fiber cement roofs and dirt floors [1] (Fig. 3.8).

AETCR Colinas "Jaime Pardo Leal"

Approximately two hours south-west of San José del Guaviare, is located the AETCR Colinas, Jaime Pardo Leal, named in honor of the first presidential candidate of the political party, Unión Patriótica. With 71 hectares of extension, it has become a rural center for the population on the edge of the jungle. It was constituted in 2017, like the other ETCRs, as a transitional space for reincorporation with around 100 inhabitants, but its advantageous connection to basic and urban services and orderly planning, which has currently attracted more than 485 people from surrounding areas and even non-demobilized population, has given it the category of a permanent and not seasonal settlement.

For these reason in 2019 the National Land Agency agreed to purchase the land for the final consolidation of the AETCR and is currently processing the acquisition

Fig. 3.8 Aerial view of the AETCR Marco Aurelio Buendia, Charras. By the author, 2021

of individual plots of land. It should be noted that Las Colinas is located in the so-called Peasant Reserve Zone, an instrument "aimed at eliminating and preventing the concentration or inequitable division of property, as well as regulating the occupation and use of the nation's vacant lands, giving preference in their allocation to low-income peasants" (Congreso de Colombia, [5]), which functions as a catalytic mechanism for their permanence.

AETCR's consolidated infrastructure consists of 236 homes built by the inhabitants with materials provided by the state. Each house is $24m^2$ for one person, with water, electricity, gas, weekly garbage collection and, in principle, collective sanitary units for every four houses. Despite the community's initial dissatisfaction with the materials provided, the houses have been modified and progressively show the appropriation of the site with vegetation, aesthetic transformations, and productive spaces. These constructions consolidate the general urban layout of the settlement in the absence of roads or public space.

In terms of common spaces, there is a library, several classrooms, a meeting kiosk and two dining rooms. Complementary services managed and built by the community include a daycare center, a memory house, a bakery, and a store run by the women's association. These facilities are adequate with wooden structures, masonry and fiber cement or zinc roofs [2]. Productivity based mainly on fish farming, pig farming, agro-ecological crops and the generation of artisanal concentrates aims to remain within a mutually beneficial system for self-sufficiency and self-management.

The last field trip relevant to the second methodological phase of this work was carried out in November 2021 in the village of El Capricho located south of the urban center of San José del Guaviare. This field trip was carried out in the framework of the workshop "*Casa Raíz*" organized by the Universidad de La Salle de Bogotá, with the objective of carrying out a series of participatory interventions that would allow the strengthening of local community processes.

For the specific case of this research, this visit allowed the recognition of the territorial conditions that frame this hamlet, particularly for its proximity to the AETCR Las Colinas and for being the most important population center in the south of San José del Guaviare, with a health center, a school, coverage of public services such as energy and aqueduct and a supply of local commerce that serves several neighboring hamlets (Fig. 3.9).

The field work carried out during these five visits over three years revealed relevant findings for this research and for the review of the results of the architectural projects built within the framework of the AETCRs and PDETs. First, there is a lack of knowledge on the part of the designers of the determinants and conditioning factors of the territory in the architectural design process of housing, community centers, sanitary facilities, and support equipment for productive activities. Second, the design in these scenarios did not involve the concerns, practices, and local knowledge, either individual or collective of the local inhabitants. Third, a vision of a sustainable future for the communities was not integrated. There is a provisional character of the houses and a disarticulation of the architecture with the customs and ways of living.

3.2.3 Life Stories that Give New Meaning to Space

The fieldwork experience allowed direct interaction with local people who are part of the history of the territory. Some of them have lived through conflict events such as armed attacks, massacres, and displacements. In the dynamics of the work with the communities, it was possible to talk openly with some people who shared life stories that show the nature of the Colombian armed conflict from the point of view of those who lived through it. This section presents three stories told directly by their protagonists. These stories are associated with space and the way in which it can be redefined through tragic events, but also through reparation and reconciliation.

Fig. 3.9 Aerial view of AETCR Jaime Pardo Leal, Colinas. By the author, 2021

Alejandro Rios "Colmillo" and the School of Charrasquera

In 1980, when Alejandro Ríos, known by his neighbors by the nickname "Colmillo," arrived in the jungles of Guaviare, he set up a tent with plastic sheeting and leaves so as not to sleep in the bush. He arrived looking for mountain land to plant coca. Five years later he built the first house on a hill where the cars could reach. Other settlers were encouraged to build more houses, and so the hamlet was configured and populated the village of Charrasquera. Some preferred to build in the surrounding areas. By the beginning of the 90's there were already about 30 houses in the hamlet and others within the boundaries of the village. Several grocery stores, clothing stores, discotheques, butcher shops, and they had their own slaughterhouse. "Colmillo" paid ten thousand pesos to the junta every time he slaughtered a cow and sold the meat to the community and neighboring villages.

The arrival of the FARC-EP in the territory established an alternative system of government run by the guerrillas who managed the area. The inhabitants adapted to living under the rules of the armed group in the total absence of the state. They were allowed to grow and produce coca paste under the authorization and control of the FARC-EP. They could also sell it by paying a tax to the armed group to deliver it to drug traffickers or landowners who collected it from the farms. In 1994, with a consolidated community, they built the school together and began the construction of the main road of the village with the help of the governor's office. The intention was to build a wider road on the same path that years before the guerrilla groups and the first settlers had marked, but the work was left unfinished due to the interference of the guerrillas who took away the machinery, leaving a large part of the road and the work almost finished.

Until the late 1990s, the community of Charrasquera lived in peace and enjoyed the money they earned from the cultivation and production of coca paste. They had large tracts of land, and people came from various parts of the country to work as coca harvesters or in different jobs in an economy that always revolved around coca cultivation. The settlers became accustomed to the excesses. Every weekend they would get together to celebrate, organize soccer tournaments, and engage in leisure activities, games, and gambling. They continued to build houses and the community grew substantially.

In 1998, the arrival in the region of the paramilitary group Autodefensas Unidas de Colombia (AUC) marked the beginning of a crude territorial war characterized by constant clashes between illegal armed groups. During the following years, there were frequent disappearances of peasants and forced displacements of several communities caught in the crossfire. Any slightest gesture by the inhabitants towards one of the groups was considered by the other group as support, and consequently generated retaliation. In September 2002, the paramilitaries entered Charrasquera: "that day in the school they hid the children in the bathrooms to protect them from the bullets" [15]. During the following days, the community began to divide over differences in sympathizing with one group or another. Some paramilitaries settled in the hamlet causing discomfort among the inhabitants forcing them to cooperate and provide them with different services. The community did not want to participate in the war and tried to stay out of it, but they were constantly intimidated and threatened by the armed groups.

As a result of this territorial dispute, on October 31, 2002, a group of FARC-EP guerrillas entered the hamlet while the inhabitants were celebrating Halloween. They forced the peasants to move to the soccer field and set fire to the houses one by one. In this area of the country, houses are built of wood, so all of them were quickly consumed by the fire. Only one building in the hamlet was not made of wood. It was the power plant. A small construction of six square meters in brick and concrete to which a grenade was thrown during the attack.

Another grenade was thrown on the school. The same construction that the community had built a few years before and that was a symbol for the community of the formation of a generation of children and young people, the future of the hamlet. The inhabitants of Charrasquera ran scared between the fences and the trees, moving away from the village and abandoning their territory. Doña Marina, one of the oldest inhabitants of the hamlet, was in the village on the day of the victimization. In her words she narrates what happened that day:

> That day we all left here, more than 70 people, [...] the guerrillas got involved with the paras (paramilitaries). We were celebrating the day of the little witches, they took us to the field, they told us all to stay there, they wouldn't let us go to the farm [...] Luckily, when those paramilitaries signaled us to go out, we went out that way, we knocked down a wire fence that was there, we went over and we went out and a bomb fell. If we had been in that crowd, they would have killed us on the field [4].

Alejandro was not in the village that day. He was part of a group of about 15 inhabitants who responded to a previous call from the guerrillas and were taken by them into the jungle for a few weeks to protect their life. After the burning of the hamlet, he returned to Charrasquera to look for the cattle and some of his belongings. He found nothing but the rubble of the houses, some frightened cattle and the bodies of the inhabitants who were killed or who did not manage to escape the attack. "Colmillo" then decided to leave the area and since then has been living in Guanapalo, the neighboring hamlet.

Between 2003 and 2005 the village remained in ruins. Only in that year, after the demobilization of the paramilitary group AUC, several of the farmers returned to the hamlet and to the coca crops under the control of the FARC-EP. Since 2012, with the start of peace talks with the guerrillas, more people who had been displaced in the past have been returning to recover their land and rebuild their homes. In 2016, with the signing of the Agreement, the guerrillas withdrew definitively from the territory and began a process in which the inhabitants committed to replace coca crops with other economic activities.

The ruins of the power plant and some traces of the school's foundations remain from the 2002 attack. In 2019, with an active and populated village, the community decided, with the support of La Salle University, to create a mural on the ruins and build a community garden around them. Also, during that same year, the young people of the community adapted a new soccer field over the old one and return to play whenever they can. The same place that was the scene of terror during the attack is now a symbol of reconciliation and life in peace. As for the school, there is no interest in rebuilding it. They prefer that the children go to the school in the neighboring hamlet.

Amparo and the Guanapalo Community Center

Amparo was born and raised in Guanapalo, a hamlet located 90 km northeast of the town of San José del Guaviare. Her mother arrived in the hamlet when she was 14 years old, as part of a family that came in search of opportunities in the coca bonanza. In the 1990s, her 15-year-old sister voluntarily joined the guerrillas with the promise of being able to be with her boyfriend, who had joined the group to avenge the murder of her father by the paramilitaries. Her brother disappeared in 2000 and was never heard from again.

From a restaurant that she runs in 2021, when we interviewed her, Amparo tells us part of her life and her past. The restaurant is located on the only street in the hamlet, where are located 60 houses that make up the hamlet are arranged. Right in front of the restaurant is the communal house of the village where all the community meetings are held, as well as parties, funerals, and other events.

In 1998, the hamlet was controlled by FARC-EP guerrillas. That same year, the paramilitaries entered the hamlet. They wore military fatigues, scarves on their heads and an armband with the letters AUC. They entered marking all the houses with the same initials. A man who was in the house next door to Amparo was shot in the head. They put him in a car and took him to Maku, a nearby village where the commanders of this group were located, to show him as a fallen guerrilla.

Terror gripped the population when they were summoned to a meeting in the community center. There, the group led by "Negro Tyson", whose real name was never known, but who was famous in the region for his cruelty and cold blood, made an intimidating speech about the control they would exercise from that moment on. Amparo tells us that she, as a 14-year-old girl, remembers that they ordered the community to line up and present themselves one by one in front of the commander.

Amparo and her family had been displaced from their farm several kilometers away and had taken refuge in the village. The fear of being displaced again was present and became much stronger at that meeting. In line, a mentally handicapped boy laughed. The commander angrily ordered him to shut up. At the same time, among the armed men is "El Boyaco". A man known in the region because he belonged to the guerrillas and escaped from there to change sides. He was the only person from the region. All the others were mainly from the Caribbean and Pacific coast.

"El Boyaco" had warned Diana's parents not to introduce her but to hide her. He knew she had links with the guerrillas. Her parents did not listen. Paramilitaries killed her when she showed up, right there at the center, in front of Amparo and all the people in the village. When Amparo and her family arrived for the interview, they were asked if they had family in the guerrillas, if they knew anyone in that group or if they sympathized with them. The mother's confident and calm answer allowed them to stay alive. Amparo's mother never spoke of her eldest daughter.

The intimidation lasted several days, and the assassinations became customary. Anyone who had any connection with the guerrillas was publicly executed. Amparo's mother decided to go to the town of San José with her daughter. After the demobilization of the AUC in 2005 they returned to Guanapalo and stayed there until 2020 when Amparo's mother became the only Covid-19 fatality in the village.

Since then, Amparo has been in charge of the restaurant that is located right in front of the community center where today all kinds of community events are held. The same place that for years was a symbol of terror and death, now means for the community a place of meeting and collective construction. The bright colors of its wooden enclosure with local handicraft techniques represent the joy and hope that the inhabitants refuse to lose and that they keep alive to nurture the processes that will allow them to take root in the territory and never have to leave again.

Leider Méndez and ETCR Colinas

Leider voluntarily joined the FARC-EP when he was only 14 years old. Although it is not possible to affirm that he was fully aware of such a decision at such an early age, the conditions of his environment made him see it as the best option. Within the organization he was well received and appreciated by his comrades and especially by his commander, a man whom Leider defines as "a very disciplined old man who did not want to give up his weapons and stayed in the bush" [11].

The trust he quickly gained from his commander led him to take the National Commanders Course, a special training for armed militants who had the vocation to become commanders. Leider says that he never had any political militancy before joining the guerrilla and that everything he knows he learned inside. About the course, he says that it was quite difficult since the army was chasing them by land and air. It was 16 months of studying, training, and fleeing. The Command Course that had taken place before that one had been bombed by the army and 36 guerrilla commanders in training had died.

We met Leider during our visit in August 2021, when he gave us an interview for more than an hour. He is currently the leader of the AETCR Jaime Pardo Leal, Colinas, so he is always busy and has many responsibilities. Even so, took the time to share with us his story of how he got to this position when he was only 26 years old and how this space has evolved. Las Colinas is one of the areas in the process of consolidation as a population center. It is one of the most organized in the country. It has a gymnasium, community hotel, sports fields, restaurants, school, day care center, library, sheep production projects, experiential tourism, among others. For productive projects, they also involve people who are not former combatants, generating a positive impact and recognition among the neighboring hamlets. "People are beginning to see us as leaders who represent all the peasants who do important work for the region. We really act as political subjects" [11].

After the Course, Leider was appointed director of mass organization of the first front, one of the strongest subdivisions of the guerrillas and with the greatest influence in the east of the country. This function, relevant for the internal leadership of the front, he carried out until the moment of the surrender of weapons agreed in the agreement, when he was appointed responsible for the monitoring mechanism on behalf of the guerrilla. There he shared a table with national and international military personnel, supervising the discussions on the implementation of the peace agreement. "There are military personnel from all over, but without weapons. In this process you learn a lot. It is like a school" [11]. At the end of the monitoring phase, Leider chose to go to the Colinas ETCR. Regarding this process, he says:

> When the monitoring ended, I came here. They gave me a ranch; I fixed it up as best I could and stayed here. I had already planned to go to work as an escort because there were many old men here who had a lot of power in the FARC. I started to work with a productive project, and I left with losses. Then came the process to form the ETCR cooperative, and I

was nominated as president of the cooperative. It was my turn to take over. People nominate you with a lot of votes. I lasted as president of the cooperative for two years [11].

After becoming president of the cooperative, Leider was nominated to be the coordinator of the AETCR, the highest position in the governance structure of these reincorporation spaces, which became vacant after Iván, the first coordinator, return to arms. It was there that he began the process of leadership of his community. Later, the name of coordinator was changed and the *Junta de Acción Comunal* -Community Action Board- was established so that decisions could be made collectively, and he was elected president of this collegiate body.

Leider was a guerrilla fighter for nine years, between the ages of 14 and 23. At the time of the interview in August 2021, he was 27 years old and had been president of the Junta for one and a half years. In April 2023 he still holds that position and has led the definitive process of consolidation of the population center by managing cooperation resources for the different productive projects and specific needs of the settlement. He is a natural leader who was fortunate to be present at the transitional moment of peace building and from there he is committed to it.

According to him, the reason this space was consolidated, in contrast to others that had difficulties, is because they were certain from the beginning that they were going to stay there. At the initial discussion tables, they had several achievements in the agreements, for example, that they were assigned a house for two people instead of one for eight as they proposed; one bathroom per house instead of one for every five units; 16 hectares of land instead of the eight offered; in short, they were able to guarantee conditions that would allow them to generate roots for the space based on the vision of the future they had at the time. Many people have left, many have returned, and many more have come to live with relatives. There are many children and young people. The town center feels as much or more alive than a consolidated urban center, energized by daily activities that take place in peace, in that peace that Leider and the other inhabitants of the AETCR are committed to build and maintain.

3.3 Analysis of the Architectural Projects in San José Del Guaviare

The different architectural projects implemented within the framework of the PDETs and the AETCRs have presented diverse and questionable results that do not always correspond to the objectives set forth in the peace agreement. The architectural design for peace building must consider the variables presented in the previous chapter to contribute to those objectives aimed at the good living of the communities and the definitive overcoming of the conflict. In this section we will characterize and analyze the 17 hamlets built in San José del Guaviare. using the theoretical model and its application will allow measuring the scope of these projects in relation to peace building.

3.3.1 The Community Centers

The 17 works that will be presented below are part of the 42 PDET works that have been built in rural areas of San José del Guaviare between 2017 and 2023. They were selected for this analysis because they seek social and community strengthening in the hamlets where they are located. The analysis will be presented in two phases. First, a descriptive sheet will be presented for each one of the houses in which the characteristics identified during the collection of information in the field and in official sources are evidenced. These characteristics are classified into the four variables proposed from the theoretical model: A. Form and space; B. Activity and relationships; C. Technique and relationships; D. Technique and the environment; E. Activity and relationships; F. Activity and relationships. Activity and relationships; C. Technique and technology; and D. Sense and meaning. The second phase of the analysis aims to group the works according to common characteristics in each of the four variables. Additionally, some relevant findings related to the parameters of each of the variables of the theoretical model are presented.

The dissertation document contains the descriptive sheets incorporating the information on the projects which have been coded according to the number assigned to each community center indicated in Table 3.4, prefixing the letter D to indicate its descriptive nature. The same number is retained on all sheets to indicate which is considered in each one. The analysis sheets have been coded by prefixing the letter A to indicate their analytical nature.

According to what is presented in the descriptive sheets, it is possible to highlight some relevant aspects. All the hamlets were built with the same structural system, based on reinforced concrete frames with 25 cm thick columns and beams. The enclosure walls were built in clay blocks, painted in some cases. The carpentry and roof structure are metallic, and the roof tiles are made of red or blue galvanized sheeting. This construction system is widely used in Colombia, mainly in town centers where industrialized materials are readily available and where skilled masonry and concrete labor is easily obtained. However, the environment in which these buildings are located is a jungle environment, more than 3 h away by poor roads. Consequently, an amazon architectural typology has been developed in these places, based mainly on the materials found in the area and on the traditional knowledge of the peasant settlers and Indigenous people who have populated these territories. Thus, wood has become the most common material used in structures and enclosure walls and zinc tiles the most frequent solution for roofs.

The sloped roof responds mainly to the ease of construction and the low costs of the system. In the case of San José del Guaviare, which is a highly rainy municipality with a deficit of potable water in the rural area, pitched roofs offer the possibility of implementing rainwater collection systems, which were not considered. Likewise, the rural area of San Jose del Guaviare does not have electricity coverage, but the radiation levels offer the possibility of implementing solar energy systems, which were not included either. Despite this, the centers were built and delivered with the networks ready to be connected to the public utility systems, ignoring the deficit

Table 3.4 Community centers built in San José del Guaviare. By the author based on official data from ART, 2022

PDET community centers built in San José del Guaviare					
No	Hamlet	COP value	USD value[2]	Delivery date	Families benefited
1	Unión de Buenos Aires	$ 128.810.172	$ 33.702	4/01/2019	57
2	Charras	$ 96.901.637	$ 25.354	28/02/2020	180
3	Charrasquera	$ 129.856.842	$ 33.976	15/02/2019	45
4	San Luis	$ 164.063.841	$ 42.926	5/01/2020	26
5	Las Dunas	$ 165.570.704	$ 43.320	9/01/2020	29
6	El Morro	$ 164.078.986	$ 42.930	14/01/2020	52
7	Sabanas de la fuga	$ 167.492.622	$ 43.823	26/12/2019	94
8	San Cristóbal	$ 44.641.580	$ 11.680	19/11/2018	37
9	Caño Mosco	$ 164.270.688	$ 42.980	15/01/2020	50
10	Santa Cecilia	$ 164.270.688	$ 42.980	11/01/2020	34
11	Baja Unión	$ 165.175.034	$ 43.217	7/01/2020	30
12	La Dosmil	$ 165.175.034	$ 43.217	17/01/2020	48
13	Puerto Ospina	$ 163.041.900	$ 42.659	26/12/2019	46
14	El Rosal	$ 164.078.986	$ 42.930	20/01/2020	58
15	Caño Lajas	$ 126.092.221	$ 32.991	15/01/2019	36
16	Caño Pescado	$ 165.983.589	$ 43.428	26/12/2019	42
17	Las Orquídeas	$ 164.078.986	$ 42.930	18/01/2020	42

conditions, and implementing technical solutions for an interconnected place such as a population center (Fig. 3.10).

The budget is a relevant aspect in this type of project, and, in this case, it generates some concerns on the part of the local community. While the most economical community center built in Caño Lajas had a total investment of COP$ 126, 092.221, approximately US$, 38.500, the one in Sabanas de la Fuga cost COP$ 167,492,621, approximately US$ 51,000 [1]. This increase of about 30% does not seem to be justified, considering that the constructed area is similar, the structural system is the same and the distance from the town center does not vary significantly. Additionally, according to the compliance audit report to the Territorial Renewal Agency issued by the Comptroller General's Office [6], the value of the construction contract for the Sabanas de la Fuga community center was COP$ 91,732,994, approximately US 28,000, which evidences a gap between the contracted and executed amount.

The kitchen in Colombian rural architecture is a place where a large part of the everyday time is spent, mainly for women. In this sense, the rural kitchen is spacious, ventilated, with space to place other furniture and develop activities in addition to

[2] The exchange rate considered for the value in dollars is that in effect on 01/01/2024: USD$1 = COP$ 3.822.

Fig. 3.10 Typical community centers in San José del Guaviare. By the author, 2021

the preparation of food that allow socialization and even the productivity of the inhabitants. In addition, the wood stove is the only alternative for cooking food in the absence of electricity and gas networks. The bathroom, in the traditional typology is usually a block detached from the main block of the dwelling. This makes it possible to isolate its function mainly because of tradition and ease of access to the rainwater reserve, which is collected and used without any treatment.

However, in the design of the centers, these typological characteristics were ignored and consequently the kitchens are closed spaces, with minimal dimensions and elongated proportions, equipped with a veneered countertop, a stainless-steel sink and scarce electrical outlets; and the bathroom, incorporated into the main block of the building, is perceived as alien in an environment where the modes of ways are linked mainly by domestic architecture.

These typical characteristics are taken as the basis for the analysis sheets presented in the dissertation document. The objective is to group the community centers according to similar conditions for each of the variables and to highlight some significant findings based on the relationship between the projects. This analysis will allow concluding some successes and difficulties presented with the execution of these projects built with the objective of strengthening the community fabric through meeting spaces.

First, is variable A—Form and Space with 13 of the 15 community centers built in the rural area of San José del Guaviare, these buildings present similar formal characteristics with a typology that is repeated in all cases, designed to have an open plan social area that extends outwards. The volumetry is characterized by a transversal gable roof with no overlapping along the entire length of the building. The geometric proportion of the floor plan is one to two and, in relation to the average number of inhabitants of each sidewalk, it has sufficient area for large gatherings. As for the organizing principles of the form, it is evident the presence of an axial symmetry in the longitudinal sense that makes its architectural language simple and functional.

The remaining two were grouped because of their formal variation in relation to the 13 previous ones. This corresponds to the longitudinal orientation of the roofs and the overlap between the two slopes. In addition, some present an asymmetry in the formal configuration that does not evidence an organizing principle. The continuity of the interior-exterior space is configured, by means of the open floor plan in the social area. Related to the parameters proposed for the transition, it is evident in the two groups of centers that the design did not contemplate a strategy of formal progressiveness that would allow a planned growth of the building. However, some communities have linked the social area with other pre-existing spaces that functioned as community centers.

It is also interesting in terms of the appropriation generated, some cases where extensions have been built to the original building. Two interventions coded with numbers 2 and 8 correspond to a kitchen and a bathroom respectively, which complement existing community centers. For this reason, they were not included in this analysis since their form and space are unique in the group of works reviewed.

Variable B—Activity and relationships is analyzed in the sheets by grouping them first, in 15 interventions that correspond to centers. There it could be seen that the social area has the possibility of being functionally linked to the environment by being configured from an open floor plan. The architectural program also includes a kitchen and a bathroom area. The kitchen is closed and can only be accessed through a door from the social area. The bathroom has a variation in some buildings that have a side access, while in most of them the connection is from the social area.

All the buildings are in an accessible place near a rural road that commonly requires a 4 × 4 vehicle. In terms of the contribution to the transition identified in this variable, it is evident that the centers built in this context did not consider the differences between the communities in terms of functional needs, daily activities, and ways of living. This is due to the standardized approach to design and the lack of participation of the communities in the decisions during the design phase.

The second group correspond to a complementary function to pre-existing centers: the kitchen for Charras and the bathroom for San Cristobal. In both cases, a direct functional relationship was proposed with the building as they were built adjacent to it. Likewise, both works have an interior-exterior transition space through which this functional relationship takes place. A relevant finding in this case is that the works considered spatial and functional needs different from those of other hamlets due to the active use of the original hamlet and its good state of conservation. This allowed the resources to be allocated to the improvement and not to the replacement of the community structure. The decisions to build complementary spaces to the hamlets were consulted and approved by the community.

Variable C—Technique and technology is analyzed in an integrated manner for all 17 works, considering that their characteristics in this aspect are similar. The works are comfortable in terms of climate because they have open spaces. This is appropriate for the jungle environment where they are located, characterized by a hot and humid climate. The construction system of all the buildings is concrete porches with block walls and sheet metal roof tiles. In most cases the walls were plastered and painted. Regarding the financial factor, there were cost overruns in the works, which are attributed to administrative causes and to the transportation of materials and labor brought in from outside. The main finding in relation to the design parameters for the transition in this variable was that the works do not include ecotechnologies. They have indoor plumbing and electrical networks, but their design did not consider rainwater management systems or solar energy, even though they are in areas with high solar radiation and rainfall. In addition, these are non-interconnected areas where there is no public service coverage. Another relevant aspect is that the use of local construction practices and knowledge of the local inhabitants was not considered.

Last, is variable D—Sense and Meaning. This variable was defined in the theoretical model as a new variable that can contribute to conventional design practices to contribute to the transition to peace. In the case of the 17 works, they are proposed as meeting places that have the potential to be the scene of important decisions that allow the promotion of autonomy and vision for the future. In this way, they contribute to the PDET's objective of strengthening the social and community fabric. However, the process of design and construction of the buildings did not have sufficient community participation and collaboration, which meant the loss of important opportunities not only to generate employment but also to link training practices that included local and experts' knowledge. By not allowing people the possibility of building what is their own, the appropriation of the space has been slow. The lack of participation in decision making also resulted in a lack of understanding of local needs that would allow the design practice to be correctly located (Fig. 3.11).

Some cases, such as the community centers in Unión de Buenos Aires, did not generate any sense of appropriation. In fact, the building was found in a state of abandonment and without roof tiles. Other centers, such as those in Las Dunas, Santa Cecilia and Las Orquídeas, showed a partial state of abandonment and infrequent use due to their isolated location far from populated areas. In contrast, some buildings such as Charrasquera, Sabanas de la Fuga, Caño Mosco and Caño Lajas, showed an active use and appropriation by the community for some uses such as stores, warehouses, parties, and meetings.

Finally, there is important evidence of the way in which some communities have assumed space as their own and have designed local interactions. The inhabitants of the Charrasquera hamlet built the community kitchen in the workshop described above; in the Charras and Baja Unión hamlets the walls were decorated with murals; in La Dosmil, El Rosal and Caño Pescado, the centers were enlarged with adjacent constructions to improve their capacity. These interventions made it possible to demonstrate a design of interaction based on the execution of the work and a

Fig. 3.11 Community centers under intervention. By the author, 2021

relationship with a real need of the community to conduct numerous community activities.

3.4 Partial Conclusions

The peace agreement has been a unique historic opportunity to overcome a long and painful armed conflict. In addition, it has become a scenario for the contribution of different sectors of society to the construction of supportive ways of life that promote life in peace. These scenarios include the role of design and architecture in strengthening the community fabric in the places most affected by the war, through devices such as meeting and collective construction spaces, housing improvement and the incorporation of everyday design practices that promote autonomy, a vision of the future and the appropriation of space.

In this complex task, the public policies proposed for the implementation of the agreement have facilitated spaces for action in the municipality of San José del Guaviare and other 169 municipalities prioritized in the post-agreement. In the rural area of this municipality, field work and the contrast of information with official and scientific sources allowed identifying that the architectural design of PDET works and transitional housing for the AETCR were carried out following prototypical and standardized guidelines tending to optimize the construction process and the costs of design and construction. One of the objectives of the projects under these programs is to promote community participation in planning, execution, and subsequent use. However, in the analysis and field work, it became evident that in these cases this participation was minimal. The challenge of an architectural project in the transition to peace must go beyond institutional intentionality and its central object must be the habitability and wellbeing of local communities.

The distance taken between the designer and the end user in these projects results in projects with minimal impact on community dynamics. In these cases, the needs of architecture and construction thought from the reality of urban centers must be transferred and localized to produce an understanding of the local needs of the transition to peace, and to change the paradigms of architecture that qualify it by its technical, spatial, and formal conditions. A comprehensive understanding of the environment in which an architectural project is located allows us to imagine how it will function over time and to design the interaction of the user or inhabitant with space. Design with knowledge of how things work will be qualitatively different from any kind of design that does not [17].

For this, it becomes imperative to propose social and technical management tools to generate alternatives for the implementation, design and construction of habitat, housing, equipment, and collective spaces projects [9]. In this way, the interaction that these spaces are generating in the different territories can be improved and a useful antecedent can be established for the structuring of future phases or peace-building processes. Every process involves a design, and when the design is managed collaboratively by the user, it facilitates a deep connection between the user and the designed object and the place where it is located to pursue the objectives set, which in this specific case focus on the non-repetition of a conflict that generated death, fear, displacement and the consequent loss of the sense of belonging to the space.

References

1. Agencia de Renovación del Territorio (2022) Agencia de renovación del territorio. Obtenido de Agencia de Renovación de Territorio
2. Agencia para la Reincorporación y Normalización (2021) Ruta de reincorporación social y económica. http://www.reincorporacion.gov.co/es
3. Boano C (2011) 'Violent spaces': production and reproduction of security and vulnerabilities. J Archit 16(1):37–55
4. Cerón M (2019) Unstructured interview. (Diaz Diaz F Interviewer)
5. Congreso de Colombia (1994) Ley 160, Congreso de Colombia
6. Contraloría General de la República (2020) Auditoría de cumplimiento a la agencia de renovación Del Territorio—art. Contraloría General de la República, Bogotá
7. Diaz Diaz F, Unigarro D (2022) Architecture for transition: analysis of two projects in the peace-building process in Colombia. AIP Conf Proc 2574:100005
8. Forjando P (2017) Ruta para la construcción de una paz estable y duradera. Torreblanca Agencia Gráfica, Bogotá
9. García-Reyes MI, Anzellini S (2019) Saberes compartidos del hábitat: una arquitectura para el paisaje rural. Dearq (24):34–47. https://doi.org/10.18389/dearq24.2019.03

10. Gobierno Nacional de Colombia & FARC-EP (2016) Acuerdo final para la terminación del conflicto y la construcción de una paz estable y duradera, de noviembre 24 de 2016. Mesa de Conversaciones
11. Mendez L (2021) Unstructured interview. (Diaz Diaz F Interviewer)
12. Ministerio de Agricultura y desarrollo rural (2017) Decreto 893. Bogotá
13. Molano A (1989) Siguiendo el corte. El Áncora
14. Piquard B, Swenarton M (2011) Learning from architecture and conflict. J Archit 16(1):1–13. https://doi.org/10.1080/13602365.2011.557897
15. Rios A (2019) Unstructured interview. (Diaz Diaz F Interviewer)
16. Unidad para la Atención y Reparación Integral a las Víctimas (2021) Registro Único de Víctimas. Bogotá, Bogotá, Colombia
17. Willis A-M (2006) Ontological designing-laying the ground. Des Philos Pap 4(2):69–92. https://doi.org/10.2752/144871306X13966268131514

Chapter 4
Architectural Design Tools for Transition

Abstract This chapter aims to present a set of tools that can be useful for the practice of architectural design in transitional peace-building contexts. Some of these tools have been used by expert and non-expert designers in territories with complex realities and their results have generated impacts aligned with the hypothesis of this research to the extent that they have contributed to the promotion of autonomy, the vision of the future, the appropriation of space and the parameters proposed in the theoretical model for transitional design towards peace. To present the tools, first a contextualization of the way in which the concept of tool has been understood for the purposes of this research will be made. Subsequently, a synthesis of each of the tools will be presented to demonstrate their main characteristics and approaches in relation to peace building in Colombia and the work of architectural designers within this framework.

4.1 The Tool Concept in the Specific Context

The previous chapter reviewed some of the architectural projects built in the framework of the implementation of the peace agreement in San José del Guaviare. Although many of these projects are still active and the communities are trying to generate links with the space to take advantage of them, they still perceive them as the result of a reparation or donation from the state rather than something proper, necessary, or useful. In this sense, the analysis carried out from the application of the proposed theoretical model, makes evident the need to review architectural design practices so that this link can occur from the initial phase of the project and be preserved until the future interaction with the space. "Architects, urban planners and planners for the transition are responsible for reviewing the objectives, methodologies and tools of space and habitat design to reformulate them aligned with the challenges of the transition" [2].

The theoretical model also highlights the need to transcend the variables and parameters with which architecture has conventionally been conceived based on the form—function—technique triad. To the extent that architecture involves the search

F. R. Diaz Diaz, *Architecture and Peacebuilding*, PoliTO Springer Series,
https://doi.org/10.1007/978-3-032-13585-8_4

for sense and meaning of space, the impact that projects will have on marginalized communities in the global south will be more closely linked to local needs. In this context, the designer can no longer be a peripheral actor who proposes tangential and universal design ideas disconnected from the feelings and thoughts of the people who seek to build peace daily. "Egocentric architectural statements and master planning fictions measured with the scale of the imperative are not simply misplaced; they are crimes against the future" [2]. This position suggests the need for adjustments in architectural design practices to ground them in the vision of the future, to make them participatory, collaborative, and localized, to include mechanisms for the non-repetition of conflict and to connect with the objectives agreed upon in the peace agreement.

This new epistemological and methodological approach to conceive architecture not only arises because of the review of Colombian projects in the context of peace building, but also includes the multiple transdisciplinary debates that have emerged in the Global South from the critical studies of design as expert knowledge to validate the autonomous and creative capacities of subjects and communities. The objective of empowering these capacities is based on adapting, appropriating, modifying, and transforming both inhabited space and objects to meet local needs even without the intervention, mediation, or participation of technical experts. However, architectural design should be conceived as a collaborative, transdisciplinary and transformative practice that combines the daily actions and knowledge of local communities that know through their direct experience the environment, as well as the conditions and needs that architecture for transition should consider contributing to its resolution. The application of methodological alternatives based on the full participation of the actors in these processes leads to a conscious and transformative practice of the reality in which they are immersed [4].

These design practices must begin to emerge from the grassroots, from the location of the problem, where inhabitants begin to move towards their autonomy by creating the conditions that allow for the change of norms from within or the ability to change traditions. This could involve the defense of some practices, the transformation of others and the actual invention of new practices [1]. In that defense, transformation, and invention of practices, it becomes necessary to have methodological tools that contribute to the objectives of architecture in this specific transitional process. The tools are then facilitators of the tasks foreseen in the different phases of the project, and their function is to help align the design processes and products to the variables established for the transition to peace.

4.2 The Tools

The tools have been selected after a review of design methodologies applied in Colombia and in other global contexts. They have been selected because they have a focus and utility aligned with the parameters of the theoretical model of architectural design for transition. The tools contribute to several of these parameters. In order to present the usefulness of each tool, a series of summary sheets have been prepared that form a catalogue indicating the description, objective, process, and possible products of each tool. In addition, a circular graphical theoretical model indicates which of the parameters it contributes to. Finally, an example of application is mentioned for those tools that have already been used.[1]

It is also important to note that these tools are dynamic and adaptable to the particular conditions and focus of each design practice. Likewise, they can be used by any person or community with or without the participation of expert designers. These are not the only tools available, but the ones that were identified as the most relevant and accessible for the Colombian context of peace building based on the understanding of the complex particularities of rurality, the main scenario of architectural design for the transition to peace. Tables 4.1 and 4.2 present the variables of the theoretical model and the useful tools for each parameter. These tools will be presented below.

4.2.1 T01 Corpography

The body can be understood as "a territory where the internal world and the external world intersect. We do not have a natural body but a cultural body" [12]. The body is interpreted as a scenario of relationships between the ways of inhabiting and the space where they happen, which allows the possibility of conceptualizing the experiences linked to the territory to recognize the symbolic value of memory, objects, and space through the dialogue with the body.

Trosman [12] also proposes to take "the social as the organizational: that which regulates the relations between oneself and others,and as the distributive: the way the economy, the rights and duties of citizens are distributed" [12]. In this way, the relationship between the inhabitants and the organizational scheme in which they are immersed is understood. This relationship, present in the body of everyone, can be graphed through this tool. The importance of its use lies in its potential to understand the memory of the inhabitants who have lived through decades of conflict and victimization that have conditioned the development of their ways of living.

Daily activities and knowledge can be evidenced with exercises that apply this tool to understand the value of the local and its contribution to the generation of sense and meaning of space. Upbringing, food, sexuality, education, construction practices, economic activities, religion, among many other elements that are part of the intimate relationship of each body with the territory can be graphed through this tool. In this way the reality of the world is internalized in a conscious and unconscious way, a model of the body is constructed explaining the traces present in it, in order to finally develop useful hypotheses for the design processes (Fig. 4.1).

Recognizing the body as a material to apply a tool, also generates the understanding that the body is a capital, i.e., a resource integrated to a system of equivalences [3]. This tool also allows a vindicating vision of the bodily space, where the

[1] The complete set of tools is available at https://sites.google.com/view/toolsfortransition/tools.

Table 4.1 The tools and their contribution to the variables A, B and C. By the author, 2023

Variable		Parameters for the transition		Useful tools	
Form and space	A	1	Flexibility and progressiveness	T01	Corpography
				T02	Models
				T03	Collaborative organization chart
				T08	Futures
Activity and relations	B	1	Understanding of differences and ways of living	T01	Corpography
				T02	Models
				T03	Collaborative organization chart
				T04	Collective mapping
				T05	Collective ethnography
				T06	Relatogram
				T07	Living planes
				T08	Futures
				T11	Located practices
		2	Consensual decision making	T02	Models
				T03	Collaborative organization chart
				T07	Living planes
				T08	Futures
				T10	Analograms
				T11	Located practices
Technique and technology	C	1	Ecotechnics	T02	Models
				T05	Collective ethnography
				T10	Analograms
		2	Acknowledgement and appreciation of the local	T01	Corpography
				T04	Collective mapping
				T05	Collective ethnography
				T07	Living planes
				T10	Analograms

private is also political and the body is conceived and lived as an agency, a territory. In peacebuilding processes from below, the body is understood as a powerful container of information that can be externalized and socialized to become collective input for participatory design.

Table 4.2 The tools and their contribution to the variable D. By the author, 2023

Variable		Parameters for the transition		Useful tools	
Sense and meaning	D	1	Encouraging autonomy	T05	Collective ethnography
				T07	Living planes
				T08	Futures
				T09	Feel-thinking machines
				T10	Analograms
				T11	Located practices
		2	Co-design and participation	T01	Corpography
				T02	Models
				T03	Collaborative organization chart
				T04	Collective mapping
				T05	Collective ethnography
				T06	Relatogram
				T07	Living planes
				T08	Futures
				T09	Feel-thinking machines
				T10	Analograms
				T11	Located practices
		3	Education and local knowledge dialogue	T01	Corpography
				T02	Models
				T05	Collective ethnography
				T07	Living planes
				T09	Feel-thinking machines
				T10	Analograms
				T11	Located practices
		4	Vision of future	T03	Collaborative organization chart
				T04	Collective mapping
				T05	Collective ethnography
				T06	Relatogram
				T07	Living planes
				T08	Futures

(continued)

Table 4.2 (continued)

Variable		Parameters for the transition		Useful tools	
				T09	Feel-thinking machines
				T10	Analograms
				T11	Located practices
		5	Appropriation of the space	T02	Models
				T03	Collaborative organization chart
				T04	Collective mapping
				T05	Collective ethnography
				T06	Relatogram
				T07	Living planes
				T08	Futures
				T09	Feel-thinking machines
				T10	Analograms
				T11	Located practices
		6	Localization of the design	T01	Corpography
				T04	Collective mapping
				T05	Collective ethnography
				T06	Relatogram
				T07	Living planes
				T08	Futures
				T09	Feel-thinking machines
				T10	Analograms
				T11	Located practices

The **objective** for the use of this tool is to recognize the symbolic value of memory, objects and space through the body. The typical **process** follows these steps: 1. Reflect from the relationship with the everyday, whether they are places, objects or actions and how one interacts with them. 2. Draw a figure of the human body that represents them. 3. Point out the parts of the body that cause something on each one on an emotional level, linked to the experience and finally 4. Indicate with colors or words, how that action, sensation, object or place can be related to the highlighted part of the body. The **products** can be drawings, pamphlets, everyday practices inventory, mind maps & collages.

Fig. 4.1. Corpography. From Casadentro: Saberes tradicionales de la domesticidad cotidiana en narrativas de mujer, 2022

4.2.2 *T02 Models*

Physical models have been widely used in the design and representation processes in architecture. Their potential lies in the possibility of simulating tangible aspects of the project such as materiality, structure, and finishes, as well as intangible aspects related to proportions, spatial and functional relationships, among others. Many architectural firms have based part of their design methodology on this tool. Such is the case of renowned architects such as Frank Gehry or Renzo Piano. However, when its use is removed from the intimacy of the design office and is installed in participatory scenarios located in the place of intervention, its effectiveness in relation to the user-designed space link increases.

In the specific context in which this monograph is framed, giving the community the opportunity to participate in the creation of a prototype of the space they will inhabit, is to make them co-responsible for the design decisions and consequently take ownership of the designed object from the first moment. In a workshop where the model is used to make consensual decisions based on the prototype, objectives such as formal definition, progressiveness, materiality, technical details, among others, can be incorporated.

In addition, the collective discussion around the model object is also an opportunity to strengthen community ties and decision-making strategies. Here the role of the expert designer in facilitating and guiding the activities using this tool becomes fundamental, since, based on the experience and the objective set for the design, he/she can translate the conclusions of the exercise into design decisions that represent the wishes of the community. In this verification phase, the model can also be used to validate the result. The appropriation and localization of the design are the key to the effective contribution of this tool to the specific peacebuilding context.

The **objective** for the use of this tool is to facilitate the understanding of the space or object by interacting with prototypes that define the form, function or technique of the space or object in the design process. The **process** follows these steps: 1. Sketch the idea of what you want to project 2. Join parts of some material that allows to represent planes in volume (either cardboard, sticks, etc.) 3. Intervene once you have the base of the 3D model, gradually modifying it collectively and then 4. Detail final ideas, to reach a consensus. and manage to approximate the idea to scale. The **products** can be 3D models, exhibitions, photographs and drawings (Fig. 4.2).

Fig. 4.2 Design process using models. By AGRA architects, 2016

4.2.3 T03 Collaborative Organization Chart

Once the architectural program is established with the definition of the spaces and the areas of each one, the question of spatial relationships appears and remains in all phases of the project until the materialization of the work. When these purely functional definitions are connected to the way the inhabitants live the space daily, the interaction of the user with the built space will be more natural, fluid, and permanent. In the relationship between the dining room and the kitchen, for example, there are differences between the rural and the urban, and there are even important variations depending on the region of the country, the climate, the gastronomic culture, etc.

The differences and ways of living can be understood by means of this tool to contribute to the definition of localized spatial relationships. Additionally, this tool can be used to foresee the progressivity of space by identifying possibilities for expansion consistent with the long-term vision.

Although this tool can be applied by architects from two-dimensional representations, it can also be combined with physical models to make decisions associated with space and function. In this sense, the problem of circulation and permanence, or of social and private areas in a building, become tangible from the design phase, linking the user in the project decisions.

In contexts such as rural Colombia, the scene of armed conflict for decades, architecture has been made without architects, and homes, community halls and schools have responded to the way the inhabitants live their everyday lives and the functional needs they identify and experience. An activity using this tool will be able to translate the knowledge of the inhabitants into useful information for the spatial distribution of the buildings and the positive response to future interaction.

The **objective** for the use of this tool is to understand the architectural program of a space and propose a scheme of functional relationships through a basic scheme. The **process** follows these steps: 1. Define an architectural program, 2. Determine the areas or at least proportion of each of the spaces, 3. Organize each of the spaces on a plane, uniting them according to their relationship of use, form and/or function and then, 4. Propose organization alternatives and discuss which will be the best option. The **products** can be 3D models, schemes, diagrams, table charts and architectural plans.

4.2.4 T04 Collective Mapping

The fundamental objective of mapping is to graphically represent different earth data. In this sense, mapping techniques can be varied and focused depending on the nature of the data to be collected. Collective mapping has the possibility of communicating the way a group of people relates to the territory they inhabit to analyze reality and propose interventions. "All mapping is a critical action that must weave a network of interrelationships in order to imagine and put into practice new models of production and reproduction of the day to day" [9, 10].

This tool was applied in the design phase of the interventions conducted in the Charrasquera hamlet in San José del Guaviare during the three versions of the Workshop Thinking the Veredal Territory. The collective cartography was key for the definition of the extension of the community center, the mural, and the community garden, as well as for the identification of relevant aspects for the structuring of the development plan they are working on. In the same way, the application served to generate useful inputs for this research and for some undergraduate theses in architecture that were proposed there based on the cartography generated by the community, in the absence of official cartography.

Collective mapping also has the possibility of understanding and communicating the perceptions of the tangible and intangible components of the territory so that they can be converted into design ideas. In the Colombian rural context, the deficiency of official geographic information means that local communities are the ones who contain the information related to roads, infrastructure, farm boundaries, availability of public services, etc. In this sense, exercises using collective cartography can condense and transmit this type of information and make it useful for architectural, urban, and territorial projects.

Collective mapping is social, and therefore subjective and communal. It contains individual perceptions discussed and agreed upon by a collective. Hence the difference with the rigid maps of geographic institutes that must be made following norms and standards. Although both are useful and pertinent, for the peace building processes that are generated by the inhabitants, the cartography produced by the knowledge they have about their territory will allow them to participate in the design processes and consequently be co- responsible for the tasks involved in this process.

In addition to the potential of this tool in the acquisition of community territorial information, its application allows for a dialogue of different knowledge among the inhabitants, and for those who do not yet know well some characteristic of their territory to learn about it from the exchange with the more expert or older inhabitants. Likewise, in this reciprocal relationship, the vision of the younger ones can be stimulating in the creative processes that underlie the collective cartography. This strengthening of social and community ties is the basis for autonomous and permanent peace-building processes.

The **process** follows these steps: 1. Select a map or plan to use as a base 2. Define a series of questions around the objective of the cartography. 3. Point out the milestones or places that you want to highlight. 4. Doodle on the base with colors or symbols to highlight a relationship or make a contribution that is considered important, either personally or collectively to then, 5. Share the contributions, to consolidate the ideas that represent some common factor. The **products** can be pamphlets, exhibitions, photographs, drawings, stories, mind maps, emotion maps, choremes, field diaries and videos (Fig. 4.3).

Fig. 4.3 Collective mapping activity. By the author, 2019

4.2.5 T05 Collective Ethnography

Through this tool, qualitative data can be obtained to describe and interpret the cultural aspects of the communities. It consists of a direct interaction between the designer who applies the tool and the community for whom it is designed. In this interaction, complementary qualitative instruments can be used, such as direct observation, participant observation, open interviews, and life histories. In this way, relevant information is obtained that allows describing, analyzing, and interpreting their ways of life in order to translate them into design determinants.

By focusing on the ways of living and the representation of identity from the cultural space, this tool allows identifying, recognizing, and valuing the local. It also serves to create a collective memory from individual experiences that is useful in design processes that must include localized solutions. The application of collective ethnography as a tool for architectural design provides the cultural richness of identity and values that a project requires, because they determine design and function factors, according to the requirements of the community under study [11].

This tool also makes it possible to identify community processes focused on local government. Indigenous communities in Colombia, for example, use Life Plans as planning instruments constructed in a participatory manner based on a self-diagnosis and that allow them to guide the community's objectives for a given time. Once these processes have been identified, ethnography can contribute to the articulation of design projects with these plans and to the development of specific tasks included in them.

Collective ethnography is proposed as a powerful design tool in peace-building territories because it allows the understanding of the unique reality of each territory from the memory and culture of the inhabitants who have remained. From there, it can contribute to the strengthening of the sense of belonging, the promotion of autonomy and the non-repetition of the conflict to the extent that the space designed includes the daily interests of the community and promotes the development of ways of living in peace.

The **objective** for the use of this tool is understand the social, cultural and aesthetic dynamics through the collection and documentation of images or stories that allow characterizing the identity of a community. The **process** follows these steps: 1. Define a community and/or place, 2. Encourage approach with a series of questions, in the form of informal interviews, and if possible, record them, 3. Make a record of photographs of everyday life, after or simultaneously with the rest of the phases, 4. Open a space for dialogue as you empathize and progress with the interaction and then, 5. Record the largest amount of information, either through drawings, field journals (with dates, time, place) or more formal formats, depending on what is required. The **products** can be pamphlets, exhibitions, posters, semi-structured interviews, life stories, photographs, stories, mind maps, choremes, field diaries and videos.

4.2.6 T06 Relatogram

Technology has allowed audiovisual documentation to be immediate using instruments such as cell phones or cameras. However, this tool is based on the graphic documentation of a story through drawing. This way of documenting makes there is a deeper attention to the most important details of the story that is shared and generates a graphic synthesis from listening.

The exercise of documenting graphically with drawings is complemented with the incorporation of some notes or key words of the story. In the activities in which this tool is implemented, participants should shed aesthetic prejudices about the drawing and concentrate on the essence of the story and the authenticity of the record. When the relatograms are socialized in the subsequent phase, the contrast between the story and the perception of the story opens a collective discussion that leads to consensual ideas regarding memory, dreams, and space (Fig. 4.4).

The designer's key task in the use of this tool consists of facilitating a spontaneous and natural process that will make evident the memory of the inhabitants and the feelings it generates. Subsequently, the objective is to convert this information into design criteria that stimulate the sense of appropriation of the space, but above all the valuation of the local. This tool could be useful to discover the knowledge present in the inhabitants, necessary for the management, design, and construction phases of architectural projects in the framework of peace building.

Fig. 4.4 Exercise from a relatogram in San José del Guaviare. By the author, 2019

The **objective** for the use of this tool is to document the perception of a group of people about a place or object from a series of graphically documented stories. The **process** follows these steps: 1. Select the stories that will be shared. For this, the participation of the whole group is important, 2. Listen to the story that will be shared setting an approximate time to limit the exercise, 3. Individually draw ideas based on the stories heard, accompanied by symbols and words, and finally the 4. Socialization and discussion between what is narrated and what is drawn. The **products** can be pamphlets, posters, photographs, stories, choremes, fanzines and videos.

4.2.7 T07 Living Planes

The living planes tool combines principles of cartography and collective ethnography. It focuses on the search for innovative responses respectful of local traditions based on a direct understanding of reality. In Colombia, it has been developed and applied in complex rural contexts by architect Simón Hosie, winner of Colombia's National Architecture Award in 2004, including during the years of conflict before the signing of the peace agreement.

This tool is intended to avoid the problems identified in the design practice that lacks knowledge of the site. It is an interactive tool with detailed information about the local inhabitants, their history, customs, lifestyle, and other technical components. It consists of identifying and understanding the particularities and needs of the territory through permanent participatory exercises during a considerable period of immersion of the designer in the everyday practices of a community. This immersion is a unique opportunity to build trust with the community and access verbal and non-verbal information.

The open attitude to observation and synthesis during the application of this tool is useful to transform it creatively in function of the architectural project. This is a process that takes time and should not be rushed if it is to be done correctly [13].

In this way, the tool can contribute to reduce the risk of problems due to lack of knowledge and understanding of the local reality as evidenced in some projects built during the implementation of the agreement. The methodology of this tool is sensitive to the forms of community organization and, consequently, is useful for reconciling differences and resolving conflicts during the design phases.

It bases its effectiveness on the time spent in the research phase, prioritizing sufficient knowledge of the environment before initiating the pre- project and project phases. In this way, it is possible to identify the physical-spatial, socio- economic and cultural characteristics of the territory, and additionally to value local knowledge that can be used in the project.

The living plan system also focuses on the principles of sustainability and participation as a basis for solving the deficiencies and problems described. Including these two principles make the social contribution intended with architecture more viable by guaranteeing articulation and dialogue between local realities and global dynamics. In addition, they become axes that structure the research in its different phases and that are determinant in the designs of the urban and architectural proposal [5].

The **objective** for the use of this tool is to foster dialogue and participation between experts and communities in order to obtain a more accurate diagnosis of the territory. The **process** follows these steps: 1. Carry out a sweep of information and contact agents that are linked to the territory and are clear about the possible interests and scope of the community, 2. Immerge yourself in the place of study for a considerable time that allows the everyday collective experience, 3. Keep track of the information obtained from the recognition and observation of everyday chores of the place to understand socio-spatial dynamics from experience to then, 4. Classify, synthesize and digitalize the information collected in the immersive phase. The **products** can be pamphlets, exhibitions, posters, plans, photographs, stories, mind maps, choremes, field diaries and videos (Fig. 4.5).

4.2.8 T08 Futures

The future is that unknown moment that is yet to come and that can generate fear, curiosity, uncertainty, worry, hope, and many other feelings that we experience in the present, conditioned by past experiences. The idea of the future is also conditioned by distrust in the face of some problems that seem unsolvable, such as marginality

Fig. 4.5 Living plane activity in San José del Guaviare. By the author, 2018

and armed conflict. However, history has also shown us that it is possible to overcome difficulties and open scenarios for the construction of better futures for the inhabitants. The word "future" is standardized and seems to have a common meaning. However, the meaning that each individual give to it is different. Therefore, it is important that in the first steps of the application of the tool, the individual understanding of the word is socialized, and a consensus is established in relation to the way it will be understood in the exercise.

This tool is based on reflecting on the role we can play in shaping the reality that we collectively share, from the now to what is not yet possible and does not yet exist [8]. It is useful to project a vision of a desirable future integrating the questioning of the changes required in our environment to achieve that future. During the application of the tool, a transition is made between the imagined future and the social or collective action that implies recognizing the environment and local problems (Fig. 4.6).

While the tool allows for a vision of the ideal future, to improve its efficiency, it is recommended that it also be applied to the projection of small tasks necessary to achieve that future. Rural communities in Colombia have long dreamed of living in peace. Using this tool should not only allow them to see themselves living in peace but should also be useful to identify the actions and transformations that individuals and communities will have to make to contribute to the construction of that desirable future. There, the power of seeing the future as a tool can contribute to those everyday design processes that contribute to a larger common goal.

Fig. 4.6 Future exercise result in San José del Guaviare. By the author, 2021

The **objective** for the use of this tool is to understand the context, and the role that each one fulfills within it, according to the actions that have or could generate changes, discovering them through the stories of a community. The **process** follows these steps: 1. Define what is understood as "the future", 2. Proceed to what is known as the Polak Game. To get to know different points of view, 3. Collect stories told by different people to reinforce the exploration and socialize the information 4. Form groups of 3 to 4 people and generate 10-year scenarios, based on the previous phase, record them on a sheet, through a scheme of what will happen and how it will be. (examples recorded in the Futures guide) and finally, 5. Reflect on what was obtained by groups. The **products** can be drawings, table charts, mind maps, pamphlets, life plans, choremes and fanzines.

4.2.9 T09 Feel-Thinking Machines

Feeling and thinking are human actions that often contradict each other or are separated to justify our emotional or rational actions. Some situations require an emotional reaction, from what we feel, which is normally associated with the body and the heart. At other times, in contrast, we believe we must act rationally, with the head, with the brain. This tool seeks to generate a simultaneous response that combines what we feel and what we think and, in that way, make complex interpretations of the territory and its inhabitants. "We condense the above into a kind of mathematical saying: thinking is to feeling what the mind is to the body. We could also say: the body is to the mind what feeling is to thinking" [9, 10].

The tool proposes a transit through a series of games and activities that allow us to see what goes from feeling to thinking and from thinking to feeling. What are their relationships and how, with a handful of words, the very infinite ways in which each one of us feel-think can be defined? [9, 10]. This exercise is understood from the analogy with the Moebius strip, a non-orientable surface with a single face and a single edge, which has been used as a symbol of infinity and which poses spatial dichotomies such as going up to the bottom or going out to the inside. A Moebius strip with three folds is used as a guide for the application of the tool, facing challenges related to what we feel and what we think during the journey through it. This same transit can be transferred to try to assume, in a spontaneous and even absurd way, everyday aspects of the emotions and thoughts of the inhabitants.

Each activity should be accompanied by generating questions and recommendations that allow moving from one side of the tape to the other, from the mind to the heart, from the head to the body, from feeling to thinking. In turn, the results should be debated and socialized, trying to take other paths, refute, disagree, rebound [9, 10]. This tool is not useful for reaching consensus on decisions, but it is useful for generating trust in the other. To the extent that there is a free and spontaneous expression of sentiments, the participants can see themselves in the other, -I am another- to generate empathy and strengthen the bonds of trust that peace-building processes require (Fig. 4.7).

The **objective** for the use of this tool is to generate a debate around observation and memory, with images and words that lead to feeling and thinking. The **process** follows these steps: 1. Make a list of collective words, starting with a random word. The next participant will say another that's directly related to the previous one, 2. Collect all the words in a bag and mix them up. All words are distributed randomly and equally, 3. Compose a poem with the assigned words, 4. Extract the most resonant words from the poems and classify them by colors according to an intuitive and spontaneous relationship, 5. Collect aromas, textures, objects, natural elements, etc. Presents on the spot and 6. Compose a collage using the classified words and the collected objects, and then share it. The **products** can be drawings, collages, mind maps, pamphlets, stories, poems exhibitions and posters.

4.2.10 T10 Analograms

Analograms are diagrams that serve to dynamize and systematize workshops based on the relationship of analogies of a graphic image that facilitates critical reflection on various themes and territories, by affinity or similarity with the central figure [7]. This tool, still without documented implementations, is proposed by the Argentine

Fig. 4.7 Example of feel-thinking exercise. By La Aventura de Aprender [9, 10]

laboratory Iconoclasistas. It was proposed as an alternative for the transition from face-to- face workshops to virtual workshops during the Covid-19 pandemic.

This tool has the possibility of generating meaning from the appropriation of a word and the analogical expression of it. The word fire, for example, may mean danger for some, but for others it means the warmth of home; or tree for some represents a resource that provides fruit and wood, for others it is a common good beneficial to the environment, and for others it is a problem for agricultural activity.

To that extent, the designer who incorporates this tool in his design activities should direct it towards the generation of meaning and sense of space from the identification of the feelings and ideals of the inhabitants. As explained in chapter two,

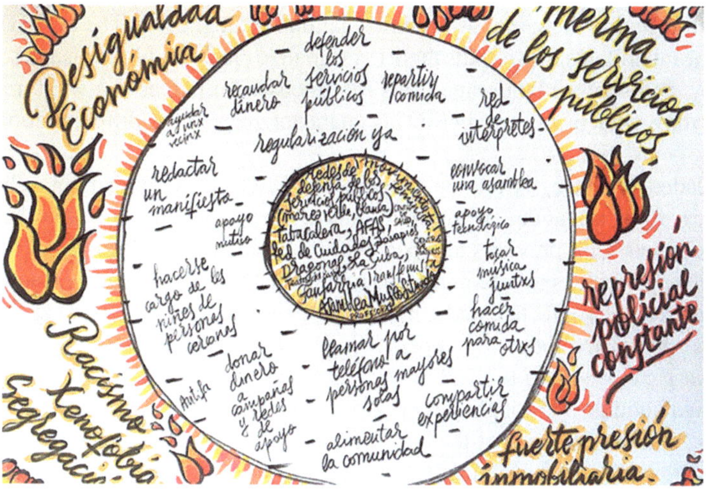

Fig. 4.8 Example of analogram. By [6]

architecture in the context of peace building in Colombia is not only an aesthetic problem but also a semiotic and political one. This tool is useful for working collaboratively on this variable and the parameters that determine it (Fig. 4.8).

The **objective** for the use of this tool is to revitalize meeting and dialogue spaces through new alternatives. The **process** follows these steps: 1. Make a group call to participate in the workshop, 2. Define the central theme and identify the keywords with the help of images that contextualize or open debate, 3. Write words related to the selected keyword individually, 4. Comment on the ideas that arise as the analogram is assembled, and finally note that the words, drawings or ideas that will form the analogy must have some kind of resemblance to the initial word or theme. The **products** can be maps, drawings, stories and photographs.

4.2.11 T11 Located practices

This tool, also proposed by Iconoclasistas, proposes to analyze the space as if it were a semiotic text. By conceptualizing it as a story, it enables a form of communication that unfolds from an ecosystem of meanings where participants, as readers immersed in this visual environment, interact, and identify this constellation of signs, enabling senses that intervene in the construction of subjectivities [7].

Although the tool is intended for application in the city, its potential in the semiotic construction of architecture makes it also useful in the generation of meaning in rural spaces in areas where peace is being built. By understanding the territory as a living story, an ecosystem of senses unfolds in which the designer is an immersed and dynamic reader capable of seeing details that are not perceived with the naked eye and translating them into useful information for the project. The everyday life in which the inhabitants move leads them to normalize elements of the visual, sonorous, olfactory, and sensory landscape. This tool focuses on perceiving what transcends the everyday, what is present silently but characterizes a place, differentiating it from others.

As readers of space, we unravel the social and cultural complexity in its historical dynamics, from a staging of the body and its sensory and perceptual apparatus, which interacts with previous and personal knowledge and experiences [6]. In this way this tool manages to interpret signs to translate them into spaces. Knowledge is generated from the affective and emotional relationship with the context. Likewise, we experiment with other ways of perceiving, changing the angle where we normally direct our gaze, smelling things that we would not normally smell or concentrating on strange sounds. "Through perception we organize the world and construct its meaning, and the presence of the body is total, even if there is no full awareness of it" [6].

The **objective** for the use of this tool is to perceive the territory sensorially through observation and experience. The **process** follows these steps: 1. Form a determined group of people, 2. Plan a tour within the territory, designing generating questions that motivate observation during the tour, 3. Start a walk with a purpose of deep and active observation trying to pay attention to unusual elements of the environment, and finally 4. Record, whether photographically, cartographically, visually, or written, answering questions that allow you to stop and observe. The **products** can be videos, pamphlets, poems, photographs, maps, plans, field diaries, fanzines, stories and drawings (Fig. 4.9).

4.3 Partial Conclusions

In chapter three, the minimal impact of the projects built in the implementation of the peace agreement on the daily community dynamics became evident. In this sense, it is necessary to review the design practices that lead to the production of these projects to reduce the distance between the designer and the user. To this end, the tools presented in this chapter can be useful as far as they focus on understanding local characteristics and needs. In the same way they allow to focus primarily on the generation of meaning and significance rather than on the technical, spatial, and formal decisions of the project. The tools then help to comprehensively understand the environment in which an architectural project is located, making it possible to imagine how it will function over time and to design the interaction of the user or inhabitant with the space.

The package of tools that has just been presented also offers the possibility of being adapted to particular needs as far as they are dynamic and versatile tools. This feature is relevant considering the differences between all territories, which makes each inhabitant and each place a universe. However, the tools must be used by designers who are sensitive to the environment and the design objective. It must

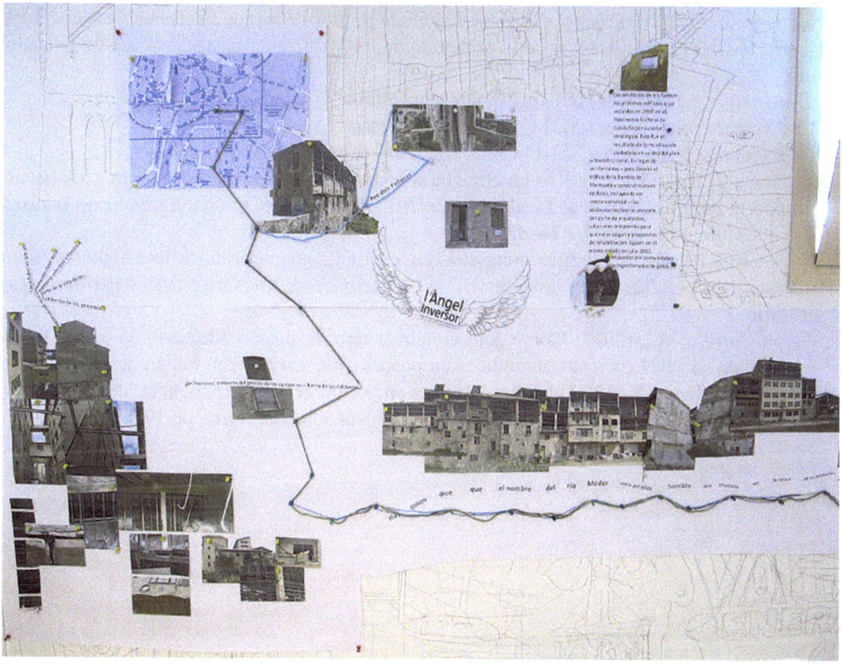

Fig. 4.9 Example of located practice result. By [6]

be understood that their application takes time that is not normally considered in the planning phases of the projects and consequently the tool itself cannot guarantee effectiveness if it is not understood and used correctly. This is the responsibility of the expert designer as a facilitator of processes that serve to make consensual decisions that remain and with which users interact. The tools must also remain as knowledge installed in the places where they were used to promote autonomy in future processes.

References

1. Escobar A (2019) Autonomía y diseño: la realización de lo comunal. Editorial Universidad del Cauca, Popayán
2. Fry T (2015) Design: on the question of "the imperative." Des Cult 7(3):417–422. https://doi.org/10.1080/17547075.2015.1105713
3. Ganter R, Daniela C, Pablo P (2018) Corpografías juveniles y generaciones en contextos de incertidumbre: subjetivación, divergencia e industria de la felicidad en el Chile actual. Última Déc 26(49):59–100
4. Geilfus F (2002) 80 Herramientas para el desarrollo participativo. Diagnóstico, planificación, monitoreo y evaluación. Instituto Interamericano de cooperación para la agricultura
5. Hosie S (2017) Planos vivos Pescaíto. Investigación participativa y desarrollo sostenible. Impresol Ediciones

6. Iconoclasistas (2020) Prácticas situadas. Obtenido de. https://iconoclasistas.net/leer-la-ciudad/
7. Iconoclasistas (2022) Analogramas. Obtenido de. https://iconoclasistas.net/cadena-de-analogias/
8. La Aventura de aprender (2012) ¿Cómo hacer futuros? España
9. La Aventura de aprender (2021a) ¿Cómo sentipensar? España
10. La Aventura de aprender (2021b) Cómo hacer un mapeo colectivo. España
11. Quintana E, Hilarraza L (2016) La etnografía como herramienta de diseño para elaborar una propuesta de arquitectónica: Ingakunapa Iachaikudiru Wasi (Casa de Aprendizaje Inga) en Cúcuta, Colombia. dearq 19:74–85
12. Trosman C (2012) Corpografías. Actas del I encuentro latinoamericano de investigadores sobre cuerpos y corporalidades en las culturas. Investigaciones en Artes Escénicas y Performáticas, Rosario
13. Varon-Garrido MA (2017) Cómo seleccionar el reto de diseño adecuado en proyectos de innovación social. Lecciones aprendidas en procesos de diseño centrado en las personas. In: Rincón ED, Tapias K (eds) Humanos a la obra en la emergencia: Experiencias de intervención de diseño para la innovación social del Caribe. Universidad del Norte, pp 107–117

Chapter 5
Conclusions

Abstract This work is the result of a research process of more than four years in which we have accompanied the rural communities of San José del Guaviare, Colombia, who have been involved in diverse ways in the armed conflict between the FARC-EP guerrilla and the National Government.

This work is the result of a research process of more than four years in which we have accompanied the rural communities of San José del Guaviare, Colombia, who have been involved in diverse ways in the armed conflict between the FARC-EP guerrilla and the National Government. During the field data collection phase, qualitative methods of empirical research were applied, such as direct observation, participant observation, open interviews, semi-structured interviews, among others. Additionally, information was recorded with instruments such as drones, cameras, GPS records and audio recorders. On the other hand, in the secondary source research, official databases and scientific sources containing qualitative and quantitative information on the different dimensions of the territories that were the scene of the armed conflict were consulted.

Simultaneously to the data collection work, the theoretical model was structured based on the contextualization of the Colombian armed conflict, the review of theories and practices associated with the design for transition, and the analysis of international references. The proposed model gathers and reinforces the conventional variables with which architecture is defined and proposes a new variable that should be prioritized in particular contexts such as those studied in this research. The contrast and analysis of the information and the application of the model produced results and contributions that have been presented in this document, structured in the three phases of methodological development described above.

In the first place, the emerging responsibility of architecture in the face of the processes of ecological and social transition in the world implies rethinking the variables with which projects are traditionally conceived. Thus, the conception of the architectural project that arises from spatial, functional, and technical problems must be complemented to think also from the social and political responsibility of architecture that give sense and meaning to space. Architectural design for transitions

F. R. Diaz Diaz, *Architecture and Peacebuilding*, PoliTO Springer Series,
https://doi.org/10.1007/978-3-032-13585-8_5

must then be conceived from practices based more on the integral recognition of the environment, the inhabitants' ways of living, local knowledge and the dreams and purposes of the inhabitants in the midst of the complex scenario of overcoming an armed conflict.

This new epistemological approach to design proposes a highly collaborative and transdisciplinary approach to generate alternatives from the understanding of the ruptures in problematic contexts. For this, the roles of designers and users must be rethought and placed in a horizontal relationship that allows understanding the creative capacities to design and solve different situations of the everyday life of the inhabitants. In the marginalized and remote territories where the conflict has taken place, the practices developed from below are more transcendental and permanent than the public policies that pretend to transform reality without knowing it. It is there where the concept of architectural design for transition becomes a practice that creates spaces to live, work and meet, contributing to the non-repetition of the conflict through the promotion of autonomy, good living, and community life.

The theoretical model developed and presented in the second chapter supports this approach by reviewing other theories and associated practices in the global south and in other relevant contexts. It also includes aspects characteristic of the context in which the armed conflict in Colombia developed that made it unique in the world. To that extent, the theoretical model is useful for reviewing, evaluating, and proposing architectural projects that seek to contribute to peace building, not only from what was agreed in the Final Agreement but also from the grassroots initiatives that will continue to emerge as communities strengthen their autonomy and their vision of living in peace. The theoretical model is also an important reference for other similar processes that may occur in international contexts of local ecological or social crises associated with the ruptures of the different types of conflict.

Secondly, the analysis of the projects in the case study reinforces the hypothesis that architectural projects that aim to contribute to peace building in a context such as the Colombian context must include localized design practices and the active and permanent participation of the beneficiary community. The historic opportunity represented by the peace agreement has generated unique scenarios for the construction of supportive ways of life that promote life in peace. These scenarios include the role of design and architecture in strengthening the community fabric in the places most affected by the war, based on devices such as meeting spaces and collective construction, housing improvement and the incorporation of everyday design practices that include the new parameters proposed in the theoretical model.

In this phase, some architectural solutions proposed for the case study and their grouping contexts were analyzed based on the proposed theoretical model, characterizing the conventional variables used in the design exercise and emphasizing the new variable, sense and meaning. Within this framework of action, it was possible to conclude that prototypical and standardized solutions have been found to optimize the construction process and the costs of design and construction. The application of the theoretical model to the architectural projects in the case study allowed us to verify the hypothesis that in the cases where the projects did not generate the results proposed in the PDET focused on strengthening social and community ties, the participation of the community in the decisions and executions was minimal. The search for appropriation of space, the promotion of autonomy and the vision of the future should be authentic searches of the communities that designers and expert architects accompany and facilitate.

The challenge of an architectural project in the transition to peace must overcome institutional intentionality and its central object must be the habitability and welfare needs of local communities, and in that sense, understand that the knowledge of these needs resides in the inhabitants. Every process implies a design, and when the design is collaboratively managed by the user, it facilitates a deep connection between the user and the designed object and the place where it is located to pursue the proposed objectives, which in this specific case are focused on the non-repetition of a conflict that generated death, fear, displacement and the consequent loss of the sense of belonging to the space.

Finally, it can be concluded that, in order to contribute to the construction of peace from the exercise of architecture, dynamic and adaptable tools are required to facilitate design practices and contribute to the objectives set after the signing and implementation of the Peace Agreement. The tools presented in the fourth chapter can be useful to the extent that they focus on the understanding of local characteristics and needs and on the generation of meaning and significance corresponding to the new design variable that the monograph presents for the transitional scenario of peace building. The Colombian conflict was a rural conflict present in most of the extension of the national territory, and each of those places present particular characteristics that make it different and unique. Consequently, designers who wish to act in these contexts will have the possibility of adapting the tools proposed in this monograph to the requirements of the project, they will be able to use them collaboratively with the user community and will be able to generate contributions that improve their effectiveness in each proposed design objective.

This package of tools also has the possibility of being transmitted to the communities so that they remain as an installed knowledge that can be used in self-managed processes of autonomous design. Although rural communities have traditionally built their spaces without the presence of architects, this contribution of expert design to local empirical practices may contribute to consensual decision- making exercises that remain. Interaction, both in domestic and collective spaces, should incorporate methodologies that involve community participation, local knowledge, and a vision of the future oriented towards the definitive overcoming of the conflict and its non-repetition.

Transfer and Continuity of Research. The results of the research can be useful as a reference document for researchers and professionals of design, urbanism, and architecture. Its practical application will favor the re-signification of architectural design in the development of an effective transition from war to peace in the Colombian context, as well as its use as a theoretical reference for other types of transitions involving vulnerable communities in contexts of social and political exclusion. For this, it is necessary to disseminate the results through publication in Colombian and international journals that are available to professionals and entities linked to the topic of design as a relevant device for transitions.

The transfer of these results should also take place in academic environments where architects and designers in training in the global south should be sensitized to the responsibility of the profession to react in a timely manner to the social and ecological crises inherent to the transition. It is important to recognize the historical causes of conflict in relation to space, in order to contribute to the non-repetition or mitigation of new conflicts through the strengthening of autonomous local processes accompanied and facilitated by professionals. This research is also a device for reciprocal training between students and inhabitants where the exchange of knowledge around the application of design tools becomes a fruitful scenario for learning and collaborative growth. These results should also be reviewed and included in the normative documents and other planning and design instruments from the public policies that focus on the implementation of the peace agreement, the good living, and the improvement of the quality of life in rural Colombia and other similar contexts.

The research constitutes a first approach to the problem of architectural design as a contribution to an effective transition from war to peace, offering assessments and proposals that integrate a new design variable and 11 parameters to be considered, among which stand out the promotion of autonomy, a new vision of the future, and the appropriation of space. Each of these new contributions opens a path towards a deeper and more detailed research on the requirements to be met by the new designs. Likewise, it is convenient to continue researching in other national and international territorial contexts, to specify the particularities of design in each of the different contexts and permanently update the theoretical model.

Each of the tools also generates new fields of research in the methodological area for the planning and management of design in historically vulnerable and excluded contexts. Although all the tools presented have already been proposed and developed by other designers, it is pertinent to verify and evaluate their effectiveness to adapt, adapt and appropriate them. In the same way, it is useful to extend the research to the many other tools that appear daily in collaborative design offices and self-managed territories.

The Colombian armed conflict has been one of the longest internal conflicts in the world and has left damage and traces that are difficult to overcome, and architecture and design can contribute to this. However, it is necessary to recognize that conflicts will continue to be part of our global history and, consequently, multidisciplinary research on the role in the prevention, mitigation, resolution, and recovery of these conflicts must be permanent and rigorous so that science can contribute to overcoming the ruptures inherent to these civilizational crises of transition.

MIX
Papier aus verantwortungsvollen Quellen
Paper from responsible sources
FSC® C105338

If you have any concerns about our products,
you can contact us on
ProductSafety@springernature.com

In case Publisher is established outside the EU,
the EU authorized representative is:
Springer Nature Customer Service Center GmbH
Europaplatz 3, 69115 Heidelberg, Germany

Printed by Libri Plureos GmbH
in Hamburg, Germany